D1158292

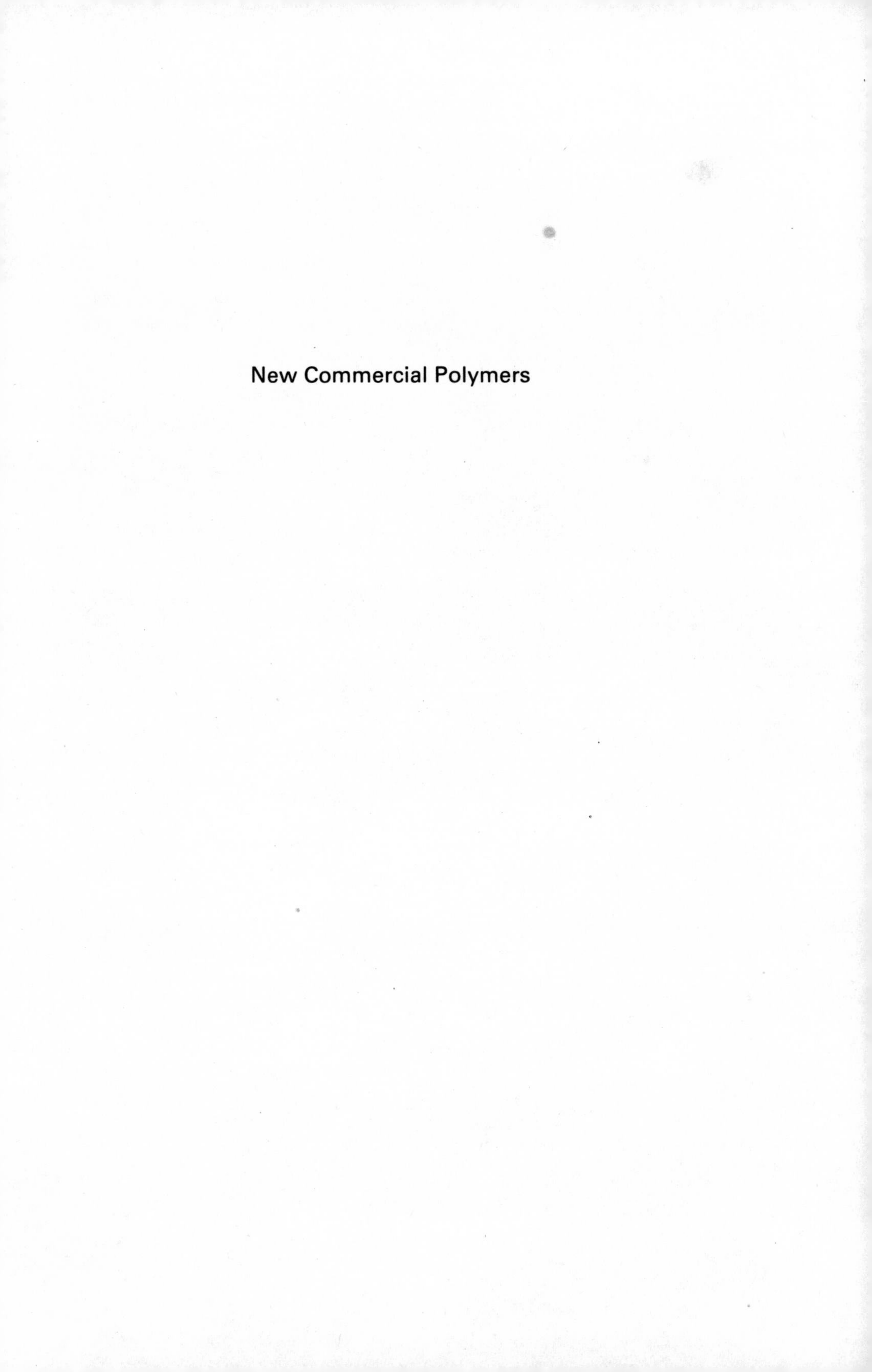

New Commercial Polymers

New Commercial Polymers 1969–1975

Hans-Georg Elias

Midland Macromolecular Institute
Midland, Michigan 48640

Translated from the German by Mary M. Exner

GORDON AND BREACH SCIENCE PUBLISHERS
New York London Paris

Gordon and Breach Science Publishers, Inc.
One Park Avenue
New York, NY10016

Gordon and Breach Science Publishers Ltd.
42 William IV Street
London WC2N 4DF

Gordon & Breach
7–9 rue Emile Dubois
Paris 75014

Library of Congress Cataloging in Publication Data
Elias, Hans-George, 1928–
New commercial polymers, 1969–1975.

Revised and updated translation of the 1975 ed. of the author's Neue polymers Werkstoffe, 1969–1974.
Bibliography: p.
Include index.
1. Polymers and polymerization. I. Title. TP1087.E4413 1977 668 76-53662

ISBN 0-677-30950-3

Filmset in Ireland by Doyle Photosetting Ltd., Tullamore, Co. Offaly

Foreword to the German Edition

Over the last few years, predictions periodically appeared that in up-coming years one could hardly expect any new classes of chemical materials in the field of synthetic macromolecules. As this book shows, these predictions have proven false, and they will continue to prove false, because they underrate human inventiveness. At one time or another the author of this book has also leaned towards the position that new plastics were scarcely to be expected and that activity would be concentrated much more on the modification and the optimization of synthetic polymers already in existence and on the working out of new fabrication processes. However, he was quickly set straight during the preparations for the third German edition of his textbook *Makromoleküle*. The manuscript of the first edition of this textbook was turned over to the publisher in 1969, that of the third edition in 1974. In this five-year period a large variety of new synthetic polymers came on the market. The author reported on these in the scope of his consulting activities for various companies. The desire of these companies to have written versions ultimately led to the concept of this book.

The present book is fundamentally a progress report which is based on references available to me or company brochures kindly sent by many firms. The book is subjective in the sense that no attempt was made to systematically and scientifically search out all relevant literature. Information concerning commercially manufactured synthetic and half-synthetic macromolecular substances is incorporated, without consideration as to their use as thermoplastics, thermosets, elastomers, fibers, coatings, synthetic leather, thickening agents and so forth. All those materials are viewed as "commercial" which either went into production or into experimental production or whose commercial sale was announced.

Naturally, the dividing line between "commercial" and "not yet commercial" cannot be sharply drawn. A few products such as Union Carbide's Bakelite Polyphenon® and DuPont's Fybex® Titanat-fibers have been withdrawn from the market in the meantime; they were not included in this book. On the other hand, the book does contain information for instance on poly(α-amino acids) and poly(parabanic acids), although these polymers are not yet on the market. However, start-up of production may be expected soon. Not enough information was available to me soon enough on other polymers, for instance the AAS-polymers of Hitachi. It is also possible that one or another new development was overlooked, because many first announcements are not easily obtainable. Finally, it is often difficult to draw a sharp line between "new" and "already known" structures. In no case was information included about start-up in production of long-known plastics, even if this plastic represented a new product for the company concerned.

A progress report such as this has to be encyclopedic for the most part, that is to say without correlation between the individual sub-chapters. Several of the new polymers have more than one use, such as an engineering material and as fiber, which precludes a classification system according to areas of application. Therefore, as a rule, the chemical structure of the chain was chosen as the basis of classification. The physical and chemical meaning of the technical terms is assumed to be understood. Consultation of a good textbook will be helpful to those less familiar with the scientific fundamentals.

All the physical units given in the literature were converted into SI-units which often caused difficulties (see Chapter 1). The appendix lists conversion factors from the old to the new units. It also contains lists of abbreviations, common names of physical quantities and the addresses of companies referred to in this book.

I wish to thank Mrs. Julia T. Lee, librarian at Midland Macromolecular Institute, for help with the literature search and those companies listed in the footnotes for reprints and company literatures. Because it is planned to periodically follow up this report, I would be grateful, furthermore, for literature references and the forwarding of press notices and company literature.

Midland, Michigan Hans-Georg Elias

Foreword to the English Edition

The English edition is basically a cover-to-cover translation of the German book, *Neue polymere Werkstoffe 1969–1974*, C. Hanser, Munich-Vienna 1975. It also includes products which came out on the market in 1975. Where new information was made available, several chapters have been amended.

It is a pleasure to acknowledge the cooperation of my former coworker, Dr. Mary Exner, who translated the book quite efficiently.

Midland, Michigan, May 1976 Hans-Georg Elias

Table of Contents

Literature references are at the end of each section

1 Introduction

1.1 DEVELOPMENT OF THE POLYMER INDUSTRY

The last two decades witnessed an enormous growth in the polymer industry. The term polymer industry includes all those industries which are involved in the production and fabrication of natural and synthetic macromolecular materials, regardless of whether these are thermoplastics, thermosets, elastomers, fibers, films, nonwoven fabrics, leather, coatings or paper.

A comparison of the world's yearly production shows that the production of natural rubber and natural fibers (cotton, wool and silk) increased about 50% each over the last two decades, whereas the production of thermoplastics, thermosets, synthetic rubber and wholly synthetic fibers increased exponentially. Specifically, in the period 1950–1970 the production of synthetic elastomers increased about ten-fold, the production of synthetic fibers increased about six-fold, and the production of thermoplastics and thermosets increased by about sixteen-fold (Figure 1-1). All indications point towards the continuation of this increase for a long time, even based on conservative growth estimates (compare Chapter 14).

A breakdown of the product groups according to single polymers or polymer families shows that, at any given time, in every product group three polymer types dominate. The production figures (Table 1-1) for the United States may be considered characteristic. The principal thermoplastics are polyolefins, poly(styrene) and poly(vinyl chloride); the principal thermosets are phenolic resins, polyurethanes and urea-formaldehyde resins; and the principal synthetic elastomers are styrene–butadiene rubber, butadiene rubber and poly(chloroprene).

By and large, these polymers are avilable at lower prices than the so-called speciality polymers, which in turn assures their strong marketing position. New polymers are produced in relatively small quantities, a fact which leads to relatively high prices. Thus it is often asked whether new polymer types can succeed at all on the market place. As historical development shows, the answer to this question is "yes," providing the polymers demonstrate substantially improved properties and can be fabricated with existing equipment. On the other hand, polymers must have extraordinary properties if a fabricator is to invest capital in new equipment. This view is supported by many examples. Poly(α-olefin) production surpassed that of the older plastics

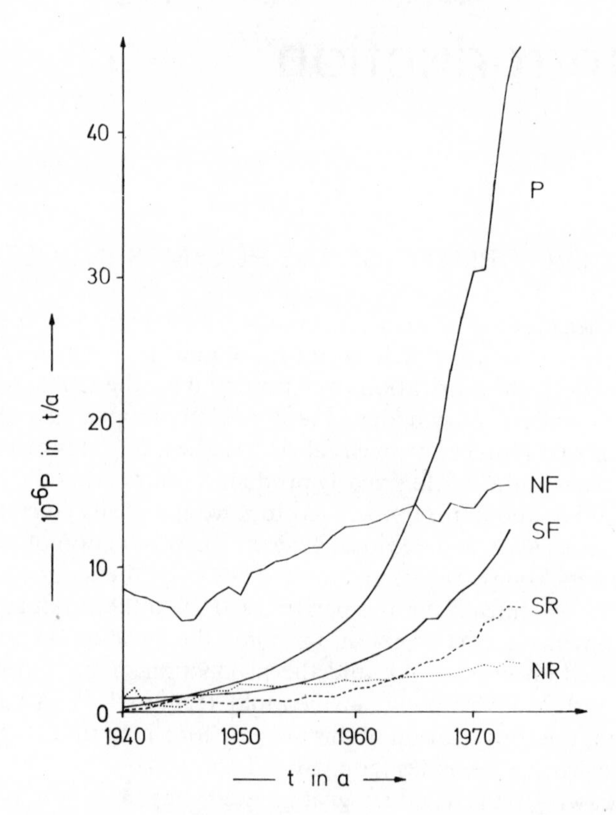

Figure 1-1 World production (in tons per year) as a function of time for thermoplastics and thermosets (P), natural fibers (NF), synthetic fibers (SF), synthetic rubber (SR) and natural rubber (NR) according to data from Refs. 1–3. The data for synthetic fibers incorporates both that for fully synthetic fibers and for cellulosic chemical fibers. The production of the latter has been approximately constant at 3.3×10^6 t/a since 1968.

poly(styrene) and poly(vinyl chloride) long ago. Terephthalic acid was still a laboratory curiosity 35 years ago, and although polyester fibers followed polyamide fibers to the market by many years, they have long since caught up with respect to quantities produced. Although polyurethanes were commercialized some 30–40 years after phenolic resins, the production gap between them has since nearly been closed (see Table 1-1).

Analysis of economic data also supports the foregoing opinion. Every economics textbook states right at the outset that there is no relation between cost and price, and the recent energy crisis dramatized this precept. Nevertheless, a relation between price and cost in a free-market economy does appear

TABLE 1-1
Production of thermoplastics, thermosets, man-made fibers and synthetic elastomers in the United States in 1970 (data of Ref. 3).

Polymer	Production t/a	
Thermoplastics		
Polyolefins (LDPE, HDPE, PP and polyolefin fibers)	2 770 000	88% of thermoplastics production
Poly(styrene) and styrene copolymers	1 610 000	
Poly(vinylchloride)	1 400 000	
Man-made fibers		
Cellulose and cellulose derivatives (*ca.* 90% as fibers; without cellophane)	810 000	79% of man-made fiber production
Polyamides (*ca.* 85% as fibers)	700 000	
Aromatic polyester fibers	670 000	
Thermosets		
Phenolic resins	460 000	67% of thermoset production
Polyurethanes	430 000	
Urea resins	410 000	
Synthetic elastomers		
Styrene/butadiene rubbers	1 352 000	79% of production of synthetic elastomers
Poly(butadiene)	284 000	
Poly(chloroprene)	148 000	

to be evident. In general, cost is increased by three factors:

raw materials
energy
environmental protection

and lowered by

technological improvements,

that is, lowered as a result of higher yields, higher reaction rates, longer catalyst lifetimes, greater production uniformity and so forth. In the past, economies from technical improvements more than offset price increases arising from climbing costs for raw materials, energy and environmental protection. Between 1950 and 1970 the prices for plastics decreased, as is shown in Figure 1-2 for the production of low density poly(ethylene), poly(vinyl chloride) and poly(styrene) in the United States.[4] Only after the abrupt price increase of crude oil did prices increase again (Figure 1-2). The increase for poly(styrene) was exceptionally large, because of the strong demand for its precursor benzene as an anti-knock agent in lead-free gasoline.

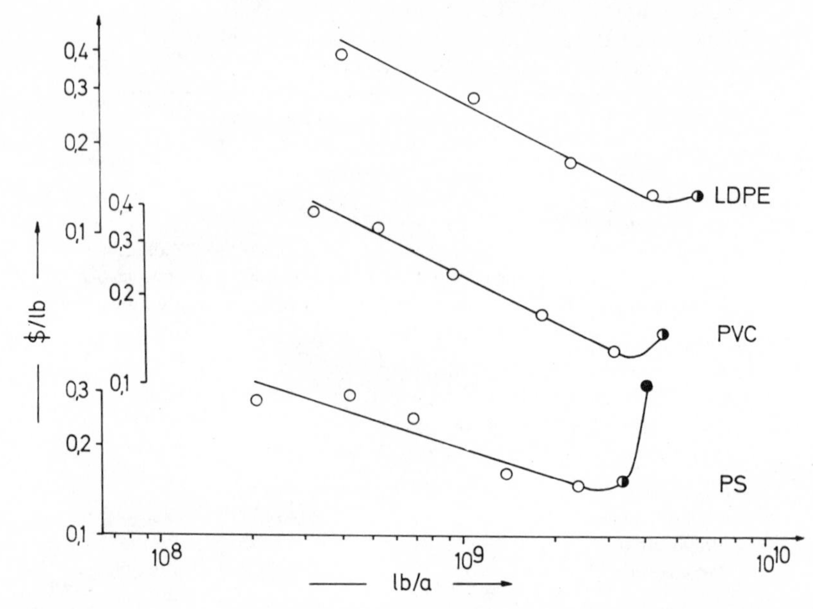

Figure 1-2 U.S.-price as a function of U.S.-production for low density poly(ethylene) (LDPE), poly(vinylchloride) (PVC) and poly(styrene) (PS) for the years 1950–1970 (○) and the years 1973 (◑) and 1974 (●). Compare Ref. 4.

The relation between the logarithm of price/unit of goods and the logarithm of production/unit of time is often termed an experience diagram. Such experience diagrams are valid not only for narrowly defined classes of materials as, for instance, certain plastics (Figure 1-2) or even all plastics[4–9] but, also surprisingly, for nearly all synthetic and natural organic and inorganic polymers over a range of nine orders of magnitude for the production year (Figure 1-3). Figure 1-3 refers to the prices and production volumes of 1969/1970. This period was chosen because it covers economically normal years. Figure 1-3 shows further that the prices for the natural polymers, wool, cotton and natural rubber, are clearly higher than those for synthetic polymers with comparable yearly productions. Even the price of crude oil ($2.50/barrel) fits on the straight line quite well; that is, it was "reasonable" if one assumes that the relation between price/volume and production/year represents a type of natural law. The price increase of about $10/barrel enacted by OPEC leads, on the other hand, to an "unreasonable" price according to this diagram.

Prices, though seldom production figures, are available for newly introduced commercial polymers for the years 1969–1974. The sparcely known data show, however, that the prices for these new polymers dovetail sur-

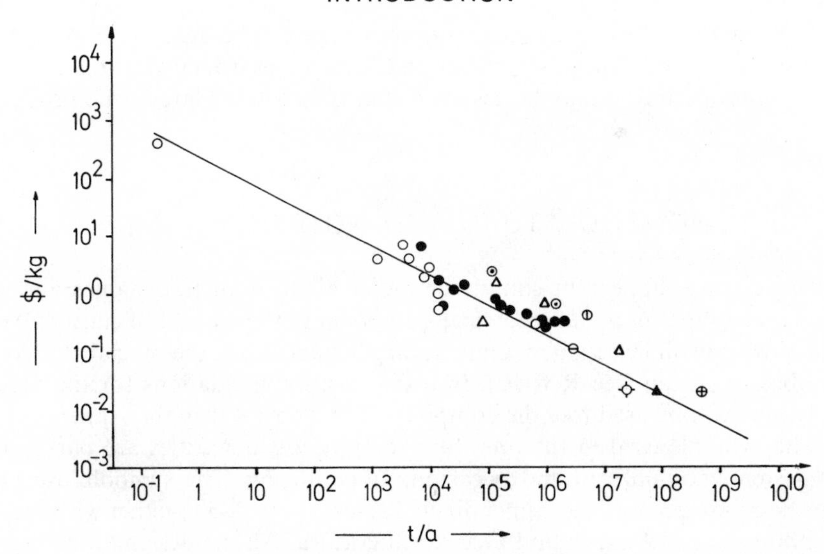

Figure 1-3 Experience diagram showing the relationship between U.S.-prices and U.S.-production or U.S.-consumption of synthetic organic polymers (●), cement (▲), vegetable gums (○), natural fibers (⊙), newsprint (⊖), crude oil (⊕), natural rubber (⦶), asphalt (⌖), and various metals (△) for the period 1969/1970. Data from Refs. 1–4, 8 and 9.

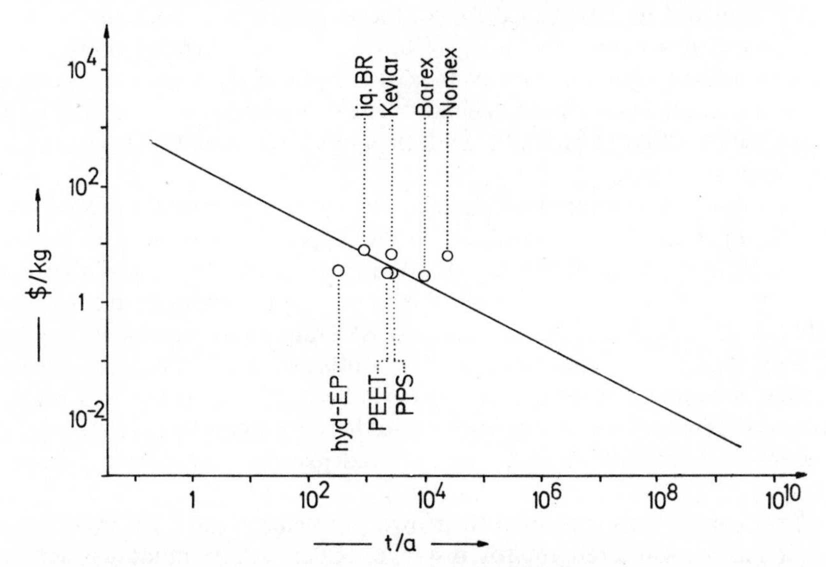

Figure 1-4 Experience diagram showing the relationship between price and yearly production for selected new commercial polymers. The line drawn is identical with that shown in Figure 1-3. hyd-EP = hydantoin-containing epoxy resins, liq. BR = liquid butadiene rubber, PEET—polyether ester, PPS—poly(phenylene sulfide).

prisingly well in the experience diagram of Figure 1-3 (Figure 1-4). As some of these polymers show significant advantages as to fabrication and/or end-use properties, it may be assumed that they will see large scale production.

1.2 NOMENCLATURE AND NOTATION

New polymers appear in commerce under all sorts of tradenames and in part under fanciful pseudochemical notations. In the interest of clarity all of these names will be retained. Only in rare instances will the official IUPAC-symbols be given (see Ref. 10). In a few cases abbreviations for the older polymers will be used (see the compilation in Table A-1 in the Appendix).

The symbols used in the literature for physical quantities are partly inconsistent, nonuniform and occasionally confusing. The symbols used in this book are given in the Appendix in Tables A-2 to A-5 together with other symbols used in English and German literature. All the designations listed in these tables were found in the literature consulted for this book, but no claim to completeness is made. With the aid of these tables it should be possible, in general, for the reader to translate favorite notations into the terminology used in this book. Hopefully a consistent international system of notation will be established for all these quantities as soon as possible. For instance it is inconsistent to say "heat distortion" when the physical unit for this quantity is temperature. It is also not exactly logical to speak of "impact strength" (with the units energy/length^2) because the word "strength" is also used for "tensile strength" with the units force/area. Further examples can be easily cited.

An additional difficulty may show up for the reader in that this book uses only SI-units. SI units are collected in the Appendix in Table A-6, prefixes for the SI units are in Table A-7, fundamental constants are in Table A-8, and conversion from old or Anglo-Saxon units into new units are found in Table A-9. This conversion to the new SI units was necessitated for two reasons. Firstly, the old units will be outlawed for commerce in some European countries after the end of 1977. Secondly, in technical literature many different units are used for one and the same quantity, so that conversion into a consistent system was required just for comparison purposes alone.

These conversions presented their own problems in part. All the cases in which mechanical strength (for instance, tensile stress, modulus, flexural strength, etc.) occurred in the traditional mass/unit area instead of force/unit area were easy to treat. More difficult was the tenacity (tensile stress, burst stress, tensile strength), which traditionally is given by mass/titer (den, tex) in the textile industry. Since titer is a length related mass, it follows that tensile strength of the textile industry has the unit of length. It can be converted to mechanical strength provided the density is known, which unfortunately is not always the case. Totally unsatisfactory is the situation with literature data on the permeabilities of liquids and gases. Conversion of the literature units (which in some cases are absent altogether) leads to not less than five different quantities, namely (see also Table A-9)

$\text{mass}\cdot\text{length}^{-1}\cdot\text{time}^{-1}$
$\text{mass}\cdot\text{length}^{-2}\cdot\text{time}^{-1}$
$\text{mass}^{-1}\cdot\text{length}^{3}\cdot\text{time}^{-1}$
$\text{mass}^{-1}\cdot\text{length}^{2}\cdot\text{time}$
time

which in turn leads one to suspect that vital information is lacking in some of these data.

References

1. Statistical Yearbook, 1972, United Nations.
2. World Almanac 1972, Newspaper Enterprise Association, New York, 1975.
3. Chemical Economics Handbook, 1974, Stanford Research Institute, Menlo Park, California.
4. H.-G. Elias, *Chem. Technol.*, **5**, (1975).
5. G. E. Fulmer, *SPE-J.*, **24**(2), 22 (1968).
6. R. Reichherzer, *Kunststoff-Rundschau*, **15**, 637 (1968).
7. G. Menges and W. Dalhoff, *Kunststoff-Rundschau*, **18**, 3 (1971).
8. N. Platzer, *Chem. Technol.*, **1**, 165 (1971).
9. R. L. Whistler, Industrial Gums, Academic Press, New York, 1973.
10. IUPAC Macromolecular Nomenclature Commission, Nomenclature of regular single-stranded organic polymers, *Macromolecules*, **6**, 149 (1973).

2 Saturated Carbon Chains

2.1 CHLORINATED POLY(ETHYLENE) ELASTOMERS

2.1.1 Structure and Synthesis

Chlorinated poly(ethylenes) have been known for a number of years. They were used almost exclusively as heat resistant plastics. The Dow Chemical Company and the Hoechst AG have now developed chlorinated poly-(ethylene) for use as elastomers (CPE Elastomers® of Dow and Hostapren® of Hoechst). The Dow CPE elastomers are available commercially as free-flowing powders with particle sizes between 300 and 500 μm. They are produced by the chlorination of aqueous suspensions of high density poly-(ethylene).[1] These amorphous polymers have in general molecular weights between 10^5 and 10^6 g/mol. The chlorine content is between 36 and 48 weight-% according to the type[2] (compare Table 2-1). Corresponding data on the Hoechst product are not available.[5]

TABLE 2-1
CPE Elastomers of The Dow Chemical Company.

Type	Chlorine content wt. %	Density g cm^{-3}	Mooney-viscosity (ML 1+4), 100°C	Limiting oxygen index, LOI %
CM0136	36	1.16	80	29
CM0236	36	1.16	40	
CM0342	42	1.25	65	32
CM0445	45	1.27	75	36
CM0548	48	1.32	80	37

2.1.2 Fabrication

The CPE elastomers can be fabricated on all conventional rubber processing equipment. Vulcanization is carried out either by organic peroxides or by irradiation. This vulcanization is a free radical process, consequently aromatics, acids, and zinc oxide must be excluded in so far as possible. Aromatics are present in mineral oil and in certain antioxidants; they act as radical scavengers. Once more, addition of antioxidants is generally unnecessary. Acids decompose peroxides ionically and thereby reduce the radical yield. Zinc oxide and, to a somewhat lesser extent, zinc stearate act as catalysts giving dehydrohalogenation. Dehydrohalogenation reactions of CPE elastomers may be prevented by the addition of lead stabilizers and epoxidated soybean oil, or by the addition of magnesium oxide.

2.1.3 Properties and Application

The thermal stability of CPE elastomers decreases with increasing chlorine content, presumably because dehydrohalogenation reactions increase also and cross-link density is reduced. On the other hand, oil resistance increases with increasing chlorine content. The saturated structure of the CPE elastomers affords them far better aging resistance than for the conventional diene elastomers. Special protection for ozone and weather stability is, therefore, generally unnecessary. Comparative studies[2, 3] on poly(chloroprene), nitrile rubbers, epichlorohydrin rubbers and chlorosulfonated poly(ethylene) showed that the aging resistance of CPE elastomers was exceeded only by that of the epichlorohydrin rubbers. The epichlorohydrin rubbers, however, demonstrate greater compression set than the CPE elastomers and are more

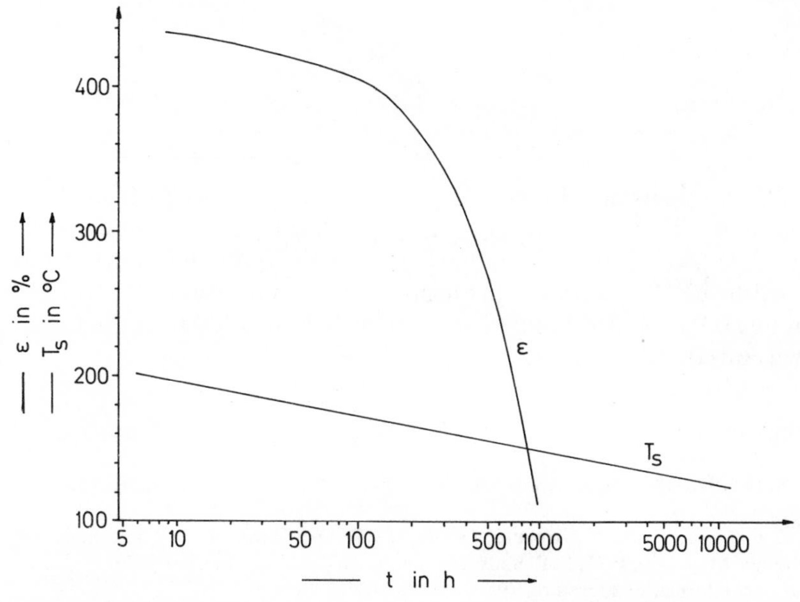

Figure 2-1 Time-dependence of elongation ε (after aging in oven at 150 °C) and of continuous-service temperatures T_S (according to data from Refs. 2 and 3.

expensive as well. When compared to poly(chloroprene), CPE elastomers show better oil and heat resistance, and when compared to nitrile rubbers, they show a better aging resistance at 150°C (see Figure 2-1).

Table 2-2 lists typical properties of vulcanized CPE elastomers. The sample mixture was comprised of 100 g CPE0136, 10 g lead silicate, 10 g epoxidized soybean oil, 15 g trialkyl mellitate, 60 g carbon black (SRF-HM-NS), 2 g triallyl cyanurate and 7.5 g dicumyl peroxide (40 % active). It was vulcanized at 160°C for 30 min.

TABLE 2-2
Properties of vulcanized CPE elastomers before and after aging in various media.

Property	Physical unit	Values measured						
		Before aging	After aging at					
			150°C oven	100°C oven	100°C ASTM oil 1	100°C ASTM oil 3	room temperature ASTM fuel B	room temperature ASTM fuel C
Modulus 100%	N/mm²	5.0						
200%	N/mm²	14.0						
Tensile stress	N/mm²	18.5	18.3	19.7	18.7	16.1	11.5	9.2
Ultimate elongation	%	325	250	325	310	250	225	200
Tear	N/cm	3.3						
Hardness (Shore A)		67	71	69				
Brittleness temperature (impact)	°C	−38						
Compression set	%		12.5					
Swelling	%				4	52	49	94

CPE elastomers can also be blended with natural rubber, SBR and EPDM. Addition of CPE to natural rubber improves the latter's oil and flame resistance while leaving the heat stability of the CPE elastomer intact. Addition of CPE to EPDM improves the former's low temperature properties while retaining oil resistance. Limited addition of SBR or EPDM to CPE facilitates its vulcanization.

References

1. U.S. Pat. 3563974; Dutch Appl. 6503760 (Sept. 27, 1965); Allied Chem. Corp; C. A. **64**, 9902f (1966).
2. L. E. Sollberger and C. B. Carpenter, lecture, Div. Rubber Chem., Amer. Chem. Soc., Toronto Canada, May 7–10, 1974; L. E. Sollberger and C. B. Carpenter, *J. Elast. Plast.*, **7**, 233 (1975).
3. Product information literature, The Dow Chemical Company.
4. R. B. Blanchard, lecture, ACS Rubber Division, Detroit, May 4, 1973.
5. Product information literature, Hoechst AG.

2.2 HIGH IMPACT POLY(VINYL CHLORIDE)

2.2.1 Structure

Rigid PVC today finds greater application as plasticized PVC. This development became obvious several years ago, consequently different companies sought by differing means to improve the unsatisfactory impact properties of rigid PVC.[1] The first impact resins used nitrile rubber, chlorinated poly(ethylene), acrylonitrile/butadiene/styrene graft copolymers or methyl methacrylate/butadiene/styrene graft copolymers as impact modifiers, principally in the form of polymer blends with PVC. The aging resistance of most of these additives is, however, not very good.

Newer developments appear to favor vinyl chloride/acrylate systems. The chemical company Lonza is offering, under the designation B 6805, a graft copolymer of vinyl chloride on dispersed poly(acrylates).[2] The composition of the poly(acrylates) is not specified, but it appears to be variable since transparency, impact strength with notch, modulus and Vicat softening temperature are optimized by the concentration, the glass temperature and the particle size of the poly(acrylate) dispersion. B 6805 contains 53.6 wt.-% chlorine and has a K value of 68 ± 1. A further development is the highly transparent EMA 5. The higher transparency was achieved by matching the refractive indices and by an inner chemical binding of the two phases. In general, EMA 5 possesses the same mechanical properties as B 6805, but it does show a somewhat lower Vicat softening temperature and a smaller flexural modulus.[2] General Tire and Rubber Co. is also offering new vinyl chloride/acrylate systems as impact modifiers for rigid PVC. Synthesis and composition of these new Esklor® systems have not yet been revealed. Not only the Lonza polymers but also the Esklor types may be blended with rigid PVC in any ratio. Lonza type B 6805 and EMA 5 as well as the Esklor type F and SR can also be used alone.

2.2.2 Properties and Application

The Lonza types represent high impact poly(vinyl chloride). The Esklor types F and SR have the properties of plasticized poly(vinyl chloride), however without the presence of low molecular weight plasticizers. Esklor type I is marketed as an impact modifier for rigid PVC, but evidently is not to be used alone. Not only the Lonza, but also the Esklor types show excellent weatherability and good mechanical properties. Stabilization is possible with all known stabilizer systems. After treatment with stabilizers, pigments and so forth, the Lonza types can be processed as a dry blend, and in fact with all equipment suitable for rigid PVC with the exception of injection molding machines. The Lonza types are employed for door and window frames and for other structural parts, such as for furniture, wall cabinets, car doors and the like.

The Esklor types require lubricants for processing, for instance poly-(ethylene) waxes. Recommended areas of application are wire and cable sheathings as well as sealants.

TABLE 2-3

Properties of impact-modified PVC-types. The property values can vary somewhat according to the stabilizer-system. They are given for the temperature of 23°C, unless otherwise noted.

		Values measured for			
Property	Physical Unit	Lonza B6805	Esklor F	Esklor SR	Esklor I
Density	g/cm³	1.36	1.20	1.25	1.20
Brittleness temperature	°C		−50	−50	−50
Vicat softening temperature	°C	80			
Tensile stress	N/mm²	50	9.9	18.6	15
Elongation	%	40	173	107	200
Modulus (100%)	N/mm²		7.5	18.2	9.2
Impact strength	N/mm²	no break			
Impact strength with notch					
23°C	kJ/m²	30			
−20°C	kJ/m²	6			
Impact tensile strength with notch	kJ/m²	250			
Tear (Graves)	N/cm		222	947	340
Flexural modulus	N/mm²	2750			
Hardness (Shore D)		78			
(Shore A)			72	94	83
Water absorption, 24 h	mg	3			
96 h	mg	10			
Limiting oxygen index (LOI)	%		22.2	24.2	

References

1. R. DeValera, *Polymer News*, **2**(1-2), 9 (1974).
2. Private communication, Dr. Th. Völker, Lonza AG, December 1974.
3. D. W. Langner, Lonza AG, lecture manuscript (1974); *Gummi-Asbest-Kunststoffe*, **28**(2), 88 (1975).
4. Product information literature of General Tire and Rubber Co.; *Plast. World*, **33** (September 22), 59 (1975).

2.3 HEAT RESISTANT POLY(VINYL CHLORIDE)

High heat distortion temperature poly(vinyl chloride) has traditionally been produced either by the chlorination of PVC, by the low-temperature polymerization of vinyl chloride or by the blending of PVC with poly(methyl methacrylate), poly(styrene), poly(acrylonitrile) or highly chlorinated polyolefins. A modified PVC, newly developed by Farbwerke Hoechst AG, Werk Gendorf, shows an increased heat distortion temperature achieved by incorporating 5–7 % N-cyclohexylmaleimide into its polymerization.

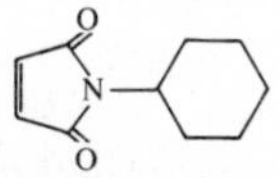

Such copolymers show a dynamic mechanical glass temperature of about 90° C (torsion bar analysis), which corresponds to a Vicat softening temperature of 87° C.[1,2] They are clear and may be processed by calendering (rigid films) and extrusion (hollow body, blow molded and injection molded pieces). Heat distortion temperature increases just about linearly with increasing comonomer content, while calenderability and deep drawability pass through a maximum at about 7 % cyclohexylmaleimide content. The initial color of uncolored copolymer is shifted more towards yellow than in pure PVC. The copolymer can be stabilized with all known stabilizer systems. Processed copolymer is somewhat more brittle than pure PVC types prepared by suspension or mass polymerization and having about the same intrinsic viscosity. The Vicat heat distortion temperature can be exceeded for short periods. The new copolymer is suitable, therefore, for applications involving brief exposure to high heat, and not especially for those involving prolonged exposure. An example of an acceptable application would be the packaging of foodstuffs while hot, such as melted cheese, jams and so forth.

References

1. G. F. Kühne, *SPE Antec*, **17**, 491 (1971).
2. G. Kühne, H. J. Andraschek and H. Huber, *Kunststoffe*, **63**, 139 (1973).

2.4 POLY(VINLY CHLORIDE)/POLY(VINYL ALCOHOL) FIBER

This matrix fiber, offered by the company Kohjiin, Japan, carries the generic name polychlal-fiber (Japan) and consists of three components: poly(vinyl chloride), poly(vinyl alcohol) and/or its acetal (presumably from chloral) as well as a graft copolymer of vinyl chloride on poly(vinyl alcohol). It is marketed worldwide under the name Cordelan® and in Germany under the name Efpakal®. For its preparation, vinyl chloride is emulsified under pressure in an aqueous solution of poly(vinyl alcohol). Polymerization leads to a mixture of poly(vinyl chloride) and a graft copolymer of vinyl chloride on poly(vinyl alcohol). The resulting polymer emulsion is mixed with a poly-(vinyl alcohol) solution and spun in a precipitating bath. From this is formed a matrix fiber of type M which contains about 50% PVC. Heat treatment and stretching leads to a dispersion of poly(vinyl alcohol) in continuous graft copolymer (type N). Acetal formation gives F-type, which is also available as shrink fiber (S-type). The M-type is used for synthetic paper, the N-type as binding fiber for nonwoven fabrics.

The F-type fibers have a kidney-shaped cross-section. They are currently available with titers between 0.17 and 6.7 tex. Their density is 1.32 g cm^{-3}. The Cordelan® fibers can be dyed with dispersion dyes, cationic dyes and sulfur dyes, but not yet with direct, acid and reactive dyes. Light fastness is good to excellent, color fastness is very good to excellent.

The Cordelan® fibers shrink only 1–2% after 60 min at 100°C. The limiting oxygen index (LOI) is 33%. The fiber does not melt on burning, rather it chars to a soft residue. Practically no toxic fumes are released on burning.

TABLE 2-4
Properties of Cordelan-fiber[1] (0.33 tex).

Property	Physical unit	Values measured dry	Values measured wet
Tenacity	km	27.2	20.3
Ultimate elongation	%	18.4	19.0
Modulus	N/mm^2	3000	
Tensile elastic recovery			
3%	%	57	
5%	%	46	
10%	%	34	
Moisture regain (65% relative humidity at 20°C)	%	3.2	

These properties predestine Cordelan® as flame resistant fiber for children's sleepwear, occupational clothing, curtains, blankets, carpeting, upholstery and wall coverings.

References

1. T. Koshiro, *Angew. Makromol. Chem.*, **40/41**, 277 (1974).
2. T. Koshiro, *Modern Textiles*, **56**, 22 (1975).

2.5 ACRYLIC ESTER THERMOSETS

Several companies have recently introduced thermosets based on acrylic esters Diacryl 101 (I) of Akzo,[1] Derakane Vinyl Esters® (II) of Dow Chemical,[2] and Spilac® (III) of Showa High Polymer Co. Ltd., Japan:[3,4]

$$\text{I}\quad CH_2{=}C(CH_3)COOCH_2CH_2O{-}C_6H_4{-}C(CH_3)_2{-}C_6H_4{-}OCH_2CH_2OOCC(CH_3){=}CH_2$$

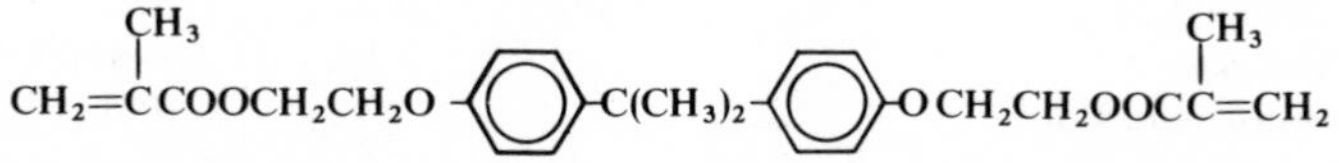

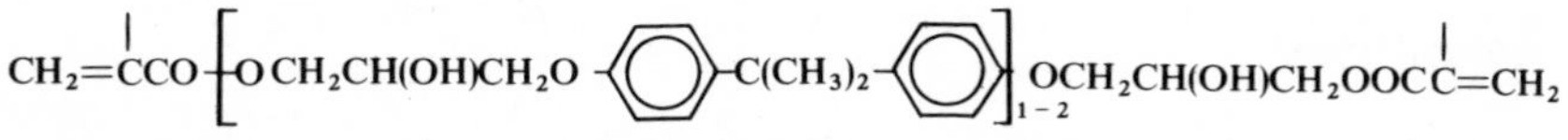

$$\text{II}\quad CH_2{=}\overset{|}{C}CO{-}\left[OCH_2CH(OH)CH_2O{-}C_6H_4{-}C(CH_3)_2{-}C_6H_4\right]_{1-2}OCH_2CH(OH)CH_2OOC\overset{|}{C}{=}CH_2$$

$$\text{III}\quad CH_2{=}\overset{|}{C}COO{-}\cdots{-}CH_2CH_2CH\begin{matrix}O{-}CH_2\\O{-}CH_2\end{matrix}C\begin{matrix}CH_2{-}O\\CH_2{-}O\end{matrix}CHCH_2CH_2{-}\cdots{-}OOCC{=}CH_2$$

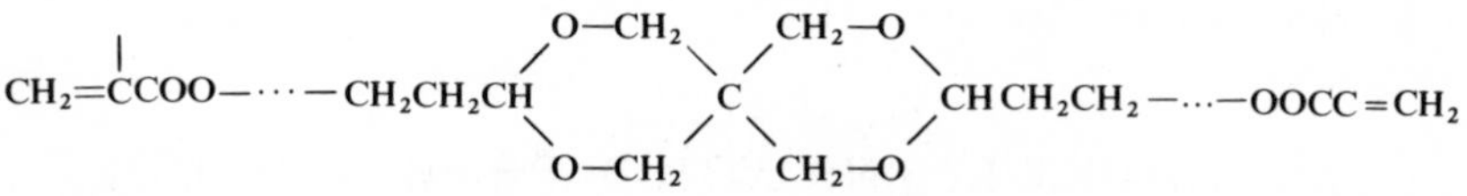

Diacryl® is manufactured by esterification of the addition product of ethylene oxide to bisphenol with methacrylic acid. The Derakane Vinyl Esters® are prepared from (meth)acrylic acid, epichlorohydrin and bisphenol A. Spilac® is produced from diallylidenepentaerythritol and (meth)acrylic acid. The Derakane Vinyl Esters® are actually mixtures of II with 40–50% styrene; this is presumably true for Diacryl® and Spilac®, too.

All three resins are thermosets and, therefore, compete with unsaturated polyesters. Shrinking at hardening is, however, lower because of the fewer double bonds *per* mass.

The Derakane Vinyl Esters® are offered in three series. The 411/412 series incorporate products for filament winding, rotation centrifuging and low pressure molding. The resins show high corrosion resistance after curing. Resins of the 510/520 series are similarly corrosion resistant and are, in addition, flame resistant. Type 470 shows increased solvent, oxidation, corrosion and heat resistance. Typical properties of cross-linked, unfilled resins are listed in Table 2-5.

TABLE 2-5

Properties of cross-linked, unfilled Derakane-vinyl esters.

Property	Physical unit	Values measured for type 411	470	510
Styrene content before curing	%	45	45	40
Tensile stress	N/mm^2	83	77	75
Ultimate elongation	%	5.0	4.0	7.0
Modulus	N/mm^2	3450	3590	
Flexural modulus	N/mm^2	127	141	127
Compressive strength	N/mm^2	117	130	120
Compressive modulus	N/mm^2	2460	2250	2320
Compressive deformation	%	6.6	11.9	7.8
Heat distortion temperature	°C	102	138	110
Hardness (Barkol)		35	40	40

References

1. H. Saechtling, *Kunststoffe* **65**, 835 (1975).
2. Product information literature of The Dow Chemical Co.
3. E. Takiyama, T. Hanyuda, and T. Sugimoto, *Jap. Plastics*, **9**(2), 6, 29 (1975).
4. E. Takiyama, *Plastics Age* **21**(9), 93 (1975).

2.6 NITRILE RESINS WITH BARRIER PROPERTIES

2.6.1 Structure and Synthesis

Nitrile resins with barrier properties are copolymers of acrylonitrile or methacrylonitrile with comonomers and which have nitrile contents of over 70%. This type of nitrile resin is particularly gas impermeable. The following four products are commercially available:

Barex 210	Vistron (Sohio) with Lonza AG as licensee
NR-16	DuPont (no longer commercial)
Lopac	Monsanto
Cycopac	Borg-Warner
PAN A-425	Solvay

In addition, American Cyanamid, Dow Chemical, ICI, Marbon, and Union Carbide are supposedly developing nitrile (?) resins with barrier properties.[1]

Barex 210 is produced by the radical graft copolymerization of 73–77 parts by weight acrylonitrile and 23–27 parts by weight methyl acrylate in the presence of 8–10 parts by weight poly(butadiene-co-acrylonitrile), in which the butadiene-copolymer contains at least 70 wt.-% butadiene[2] (compare with Ref. 3).

In the synthesis of NR-16, poly(butadiene-co-styrene) is dissolved and emulsified in a mixture of styrene and acrylonitrile. The polymerization leads to a mixture of acrylonitrile-rich styrene/acrylonitrile copolymers and their graft copolymers on the butadiene elastomer.

Lopac apparently is a copolymer of methacrylonitrile and styrene prepared by emulsion polymerization (compare Ref. 4). In contrast to Barex 210 and NR-16 it contains no elastomer. The compositions of Cycopac and PAN A-425 have not been disclosed.

2.6.2 Fabrication and Properties

The low gas permeability of poly(acrylonitrile) has been known for some time, but poly(acrylonitrile) cannot be processed thermoplastically without decomposition. The problem was, thus, to produce a thermally stable material with high acrylonitrile content. All the commercial nitrile resins with barrier properties have about a ten-fold higher gas permeability than pure poly(acrylonitrile) (Table 2-6), but still they have substantially lower gas permeability than the other polymeric packaging materials known to date.

TABLE 2-6
Gas permeabilities P of poly(acrylonitrile), nitrile resins with barrier properties and conventional packaging materials.[5]

	$10^{17} P$ in $cm^3\ s\ g^{-1}$	
Polymer	O_2	CO_2
Poly(acrylonitrile)	0.14	0.23
Nitrile resins		
Barex 210	3.6	7.2
NR-16	(4.5–9)	9–18
Lopac	2.3	4.5–9
Poly(vinylidene chloride)	3.6	14–23
Poly(vinyl chloride)	23–32	40–180
Poly(ethylene) (HDPE)	900	2000

These nitrile resins are fabricated similarly to other heat and moisture sensitive thermoplastics, that is similarly to rigid PVC (compare Ref. 6). In order to obtain clear molded objects, moisture may not exceed 0.1 %. Fabrication temperatures may not be too high. An upper fabrication temperature of 230°C is given for Barex 210 for example.[7] At too high a fabrication temperature the material turns yellow.

Mechanical properties of nitrile resins are listed in Table 2-7. Data for Cyopac have not been disclosed. Attention is called to the fact that in the reporting of gas permeabilities the values depend strongly on the pressure of the gas used (compare Ref. 6). For instance, while carbon dioxide permeability at low pressure is given the value of 5.4×10^{-17} cm^3 s g^{-1} (Table 2-7), for the same material at higher pressure the values of 17.3×10^{-17} (extruded and oriented material) and 31.6×10^{-17} cm^3 s g^{-1} (molded material) are obtained. In addition the gas permeability depends on the relative humidity of the air (Figure 2-2). All these data, however, can only be used with dif-

TABLE 2-7

Properties of nitrile resins with barrier properties at room temperature. The data listed for Lopac are taken from an article by an employee of Monsanto in which Lopac was not specifically named.

		Values measured for			
Property	Physical unit	NR-16 Refs. 6 and 9	Barex 210 Refs. 7 and 8	Lopac Ref. 4	PAN A-425 Ref. 15
Density	g/cm^3	1.12	1.15	1.15	1.17
Melt index	g/(10 min)	1–2	4		
Heat distortion temperature					
at 0.46 N/mm	°C	102	74		
at 1.86 N/mm	°C	92		105	75
Tensile yield strength	N/mm^2	74	67		
Yield	%	15	5		
Tensile strength	N/mm^2			68	84
Modulus	N/mm^2			4700	
Flexural strength	N/mm^2	120	99		
Flexural modulus	N/mm^2	3700	3400		3400
Impact strength with notch	N	54	81		
Bursting strength filled bottles[12]	N/mm^2	0.5–1	2.3		
Max. height of fall of bottles filled with 170 g[12]	cm	15–30	240		
Gas permeability					
O_2	cm^3 s g^{-1}	$(4–8) \times 10^{-17}$	3.6×10^{-17}	2.7×10^{-17}	2.3×10^{-17}
CO_2	cm^3 s g^{-1}	5.4×10^{-17}	5.0×10^{-17}	4.1×10^{-17}	3.5×10^{-17}
H_2O (90% rel. humidity, 38°C)	s		23×10^{-17}		

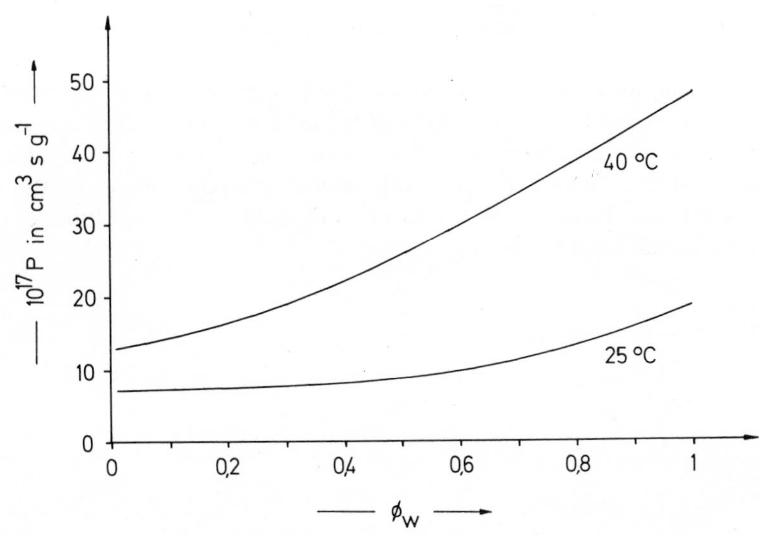

Figure 2-2 Barex 210 gas permeability to carbon dioxide as a function of relative humidity at room temperature (*ca.* 23 °C) and at 40 °C.[10]

ficulty to predict the maximum storage time of carbonated beverages in the containers from nitrile resins. Namely, the measurement of gas permeability is carried out under stationary conditions ("equilibrium" conditions), so that the initial disappearance of gas through solution in the polymer is largely ignored (Figure 2-3). In addition, the gas permeability depends strongly on

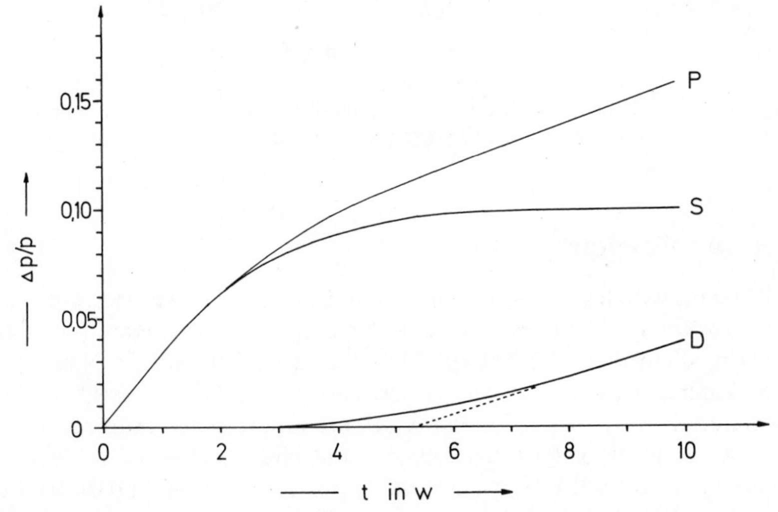

Figure 2-3 Pressure drop $\Delta p/p$ as of function of time t for the components of permeability P, solubility S and diffusion D.[10]

the orientation of the molecule chains, which in turn influences the tenacity (Figure 2-4).

Carbon dioxide loss is important for the storage of non-alcoholic carbonated beverages. For example, a 10% CO_2 loss is given for NR-16 on storage for 6 months at room temperature.[6] In the storage of beer, on the other hand, as little oxygen as possible should diffuse into the beer, otherwise the taste of the beer would be altered by oxidation. For instance, storage times of up to 2 months are advised for NR-16.

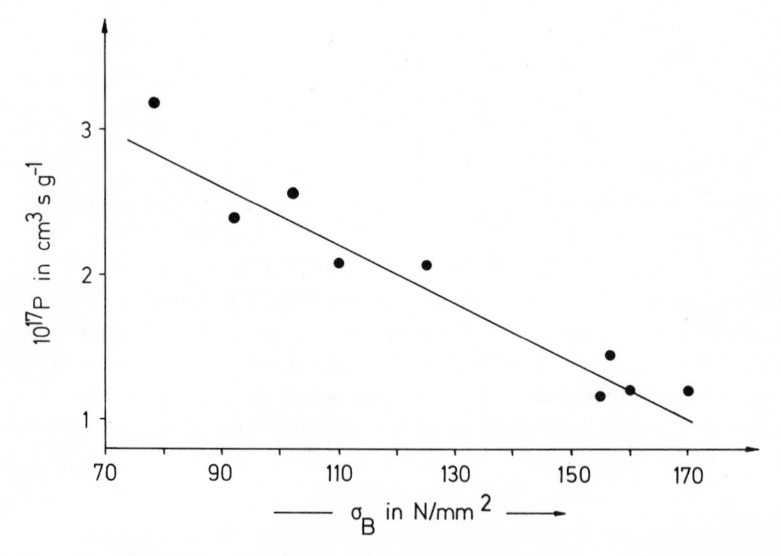

Figure 2-4 Barex 210 permeability P as a function of tensile stress, σ_B.[13] The tensile stress increases with increasing degree of orientation in the material.

2.6.3 Application

Nitrile resins with barrier properties are employed as bottles, jars, containers, etc. for carbonated and uncarbonated beverages, for cosmetics, for oils and fats, for medicines and for household and industrial chemicals. Naturally, in the packaging of foods may only as few and as nontoxic as possible extractable materials be present. For Barex 210 are reported, for example, residual monomer content of 0.01–0.05 ppm, emulsifier content of 0.1–0.5 ppm, stabilizer content of 0.1–0.5 ppm, soluble polymer content of 0.01–0.05 ppm and a chain transfer agent content of less than 0.04 ppm.[11] All nitrile resins with barrier properties appear to be suitable for food packaging.

References

1. P. Townsend, *Appl. Polymer Symp.*, **25**, 311 (1974).
2. M. T. Schuler, *Kunststoffe-Plastics* [*Solothurn*], **21**(9), 13 (1974).
3. E. C. Hughes, J. D. Idol, Jr., J. T. Duke and L. M. Wick, *J. Appl. Polymer Sci.*, **13**, 2567 (1969)
4. S. P. Nemphos and Y. C. Lee, *Appl. Polymer Symp.*, **25**, 285 (1974).
5. J. D. Idol, Jr., *Appl. Polymer Symp.*, **25**, 1 (1974).
6. H. Moncure, *Appl. Polymer Symp.*, **25**, 293 (1974).
7. K. E. Blower, N. W. Standish and R. W. Yanik, *SPE J.*, **27**(11), 32 (1971).
8. Product information literature, Vistron Corp.
9. Product information literature, DuPont.
10. D. D. Hall, Jr., *Appl. Polymer Symp.*, **25**, 301 (1974).
11. V. F. Gaylor, *Anal. Chem.*, **44**, 897A (1974).
12. Anonymous, *Plastics World*, **28**, 50 (1970)H. 5.
13. N. W. Standish, R. E. Isley and R. D. Smith, *Modern Packaging*, **41**, 45 (1974)H.1.
14. M. Th. Schuler, *Kunststoffe-Plastics* [*Solothurn*], **23**(2), 15 (1976).
15. Product information literature, Solvay, Brussels, Belgium.

2.7 POLY(METHACRYLIMIDE)

2.7.1 Structure and Synthesis

Poly(methacrylimides) (PMI) are comprised principally of methylacrylimide units.

H_3C CH_3 O N O H

They are marketed by the company Röhm under the tradename Rohacell® as rigid foam plastics.

Copolymers prepared by the radical polymerization of methacrylic acid and methacrylonitrile are used as the starting materials for the commercial production of poly(methacrylimides). Ammonia producing reagents (such as urea and ammonium hydrogen carbonate) are added to the copolymers at temperatures above the glass temperature (140°C) but under the decomposition temperature (about 240°C):

CH_3 CH_3 O=C OH C≡N ⟶ CH_3 CH_3 O N O H ⟵ ($+NH_3$, $-2\,H_2O$) CH_3 CH_3 C C O OH O OH (2-1)

The cell structure is built concurrently with the imidation. The extent of imidation has not been disclosed for the commercial product. But it depends on the base unit sequence length distribution in the starting copolymer, on the presence and type of the ammonia source, the temperature and the length of heating.[1,2]

2.7.2 Properties and Application

Poly(methacrylimides) are white, closed-cell rigid foams with high modulus, high heat distortion temperature, good solvent resistance, and good heat deformation temperature (compare Table 2-8). Flame resistant types are yellow to brown. PMI's are stable at room temperature to hydrocarbons, ketones, chlorinated solvents and 10% sulfuric acid. They are unstable to methanol, tetrahydrofuran, acetic acid and 5% sodium bicarbonate solution.

Three months at 170–210°C brought about practically no alteration of the mechanical properties. Only the modulus increased slightly, for the degree of cross-linking increased with aging. The rigid foams may be used under load to about 160°C. Creep can occur with long term loading near the glass temperature.

TABLE 2-8
Properties of Rohacell–Rigid Foams[3-5] at 20 to 23°C.

Property	Physical unit	Values measured for type		
		31	51	71
Density	kg/m^3	30	50	70
Tensile stress	N/mm^2	1.0	1.9	2.9
Compressive strength	N/mm^2	0.4	0.9	1.5
Flexural strength	N/mm^2	0.9	1.9	3.0
Shear strength	N/mm^2	0.4	0.8	1.3
Modulus	N/mm^2	30	60	100
Elastic shear modulus	N/mm^2	13	25	40
Heat distortion temperature	°C	200	195	190
Thermal conductivity	$W\ m^{-1}\ K^{-1}$	0.031	0.029	0.030
Water-diffusion resistance factor	1	400	650	900
Water absorption (98% rel. humidity, saturation)	vol.-%	0.59	0.88	1.1
Water absorption after 50 d	vol.-%	18	14	14
Dielectric constant	1	1.04	1.07	1.10
Dissipation factor	1	0.0006	0.0008	0.0010
Surface resistivity	Ω	2×10^{13}	9×10^{12}	5.5×10^{12}

PMI's are easily processed mechanically. High density types, however, are cut with difficulty with hot wires. At temperatures between 170 and 200° C the PMI's may be thermally reshaped or stamped. The rigid foams can be bonded with practically all the usual commercial adhesives. They may also be varnished.

Rohacell® rigid foams can be utilized for motor covers for buses, airplane landing gear doors, radar domes, cores for skis[6] and tennis rackets and for core layers of glass fiber reinforced materials. Through application of heat and pressure, sandwich parts with low pressure Prepregs can be prepared.

References

1. W. Ganzler, P. Huch, W. Metzer and G. Schröder, *Angew. Makromol. Chem.*, **11**, 91 (1970).
2. Brit. Patent 1045229 (October 12, 1966); Röhm and Haas; G. Schröder, W. Metzger and P. Huch, inv.; C. A. **66**, 3187 (1967).
3. W. Pip, *Kunststoffe*, **64**, 23 (1974).
4. P. Bauer, *Kunststoffe*, **60**, 546 (1970).
5. P. Bauer, *Kunststoff-Rundschau*, **17**, 165 (1970).
6. W. Pip, *Kunststoffe*, **65**, 332 (1975).

2.8 1, 2-POLY(BUTADIENE)

2.8.1 Structure and Synthesis

The polymerization of 1,3-butadiene with the catalyst system $CoHal_2$–Ligand–$A1R_3$–H_2O in solution[1-3] leads to a poly(butadiene) with about 91% 1,2- and 9% *cis*-1,4-structures; 51–66% of the 1,2 units are in syndiotactic and 49–34% are in heterotactic triades.[1] The molecular weights amount to several hundred thousand, the polymolecularity indices M_w/M_n are about 1.7–2.6. The product shows no long-chain branching. A product of this type is produced and marketed by the company Japan Synthetic Rubber under the designation JSR 1,2 PBD.

2.8.2 Properties and Application

The JSR-procedure gives 1,2-poly(butadienes) with a not too high steric order, by means of which the degree of crystallinity and thus the melting point (90° C) are held down. The glass temperature is linearly dependent on the degree of crystallinity and amounts to about —23°C for the manufactured product with a degree of crystallinity of 25%. The Vicat softening temperature is 60° C. Mechanical properties fall approximately between those of typical elastomers and typical thermoplastics (Table 2-9).

TABLE 2-9

Properties of JSR 1,2-PBD with a degree of crystallinity of 25%.[1,4,5] The samples were either formed (F), extruded through a T-shaped die (E), or biaxially oriented (0). MD = machine-direction, TD = transverse direction.

Property	Physical unit	Values measured for F	E	O
Film thickness	μm		50	30
Density	g cm^{-3}	0.907	0.907	0.907
Melting temperature	°C		90	90
Vicat softening temperature	°C		60	60
Glass transition temperature	°C		−23	−23
Tensile stress, MD	N/mm^2	112	180	40
TD	N/mm^2	112	170	530
Ultimate elongation, MD	%	415	500	390
TD	%	415	710	230
Flexural strength	N/mm^2	47		
Flexural modulus	N/mm^2	560		
Impact strength with notch		no break		
Tear, MD	N/mm		140	28
TD	N/mm		170	22
Transparency	%		91	91
Gas permeability				
carbon dioxide	cm^3 s g^{-1}		$35{,}700 \times 10^{-17}$	$34{,}500 \times 10^{-17}$
oxygen	cm^3 s g^{-1}		8200×10^{-17}	8000×10^{-17}
ethylene oxide	cm^3 s g^{-1}		$370{,}000 \times 10^{-17}$	
water vapor	s		127×10^{-17}	110×10^{-17}
Dielectric constant (1 kHz)		2.7–2.9		
Volume resistivity	Ω cm	10^{16}		
Surface resistivity	Ω	10^{16}		
Dielectric strength	kV/cm	940–1060		
Dissipation factor (1 kHz)		0.0031		

The properties of 1,2-poly(butadienes) are largely determined by the degree of crystallinity and this, in turn, by the tacticity. Products with too high a degree of crystallinity show only a small elongation and a reduced tensile stress (Figure 2-5). Product with a 25% degree of crystallinity represents, therefore, a good compromise. 1,2-PBD behaves rheologically about like low density poly(ethylene). Transparent bottles and films which are suitable for food packaging can be made from it because of its nontoxicity. The gas permeability of such films is quite high (see Table 2-9); consequently, thin films are suitable for the packaging of fresh fruit, vegetables, fish and so forth. Such foods can be sterilized with ethylene oxide in packaging of 1,2-PBD. Biaxially stretched films show increased strength and can be shrunk thermally (40% in 10 s at 80°C). Sharp objects produce no holes in films from 1,2-PBD

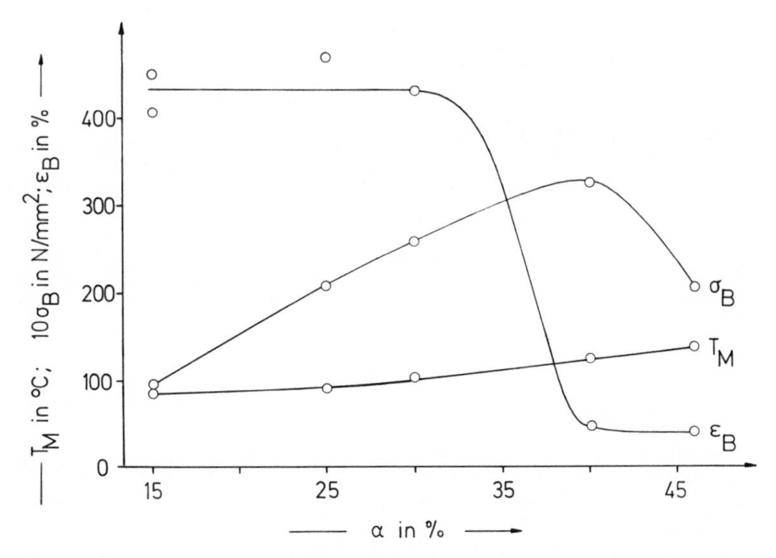

Figure 2-5 Dependence of the melting point, T_M, the tensile stress, σ_B, and the ultimate elongation, ε_B, on the degree of crystallinity a of samples of approximately equal molecular weight ($[\eta]=200$–240 cm^3/g in toluene at 30°C) (according to data in Refs. 4 and 5).

because of its high ultimate elongation; rather the film is merely stretched. Gas permeability drops sharply with increasing film thickness and reaches at film thicknesses of several mm about that of low density poly(ethylene).

1,2-PBD has many reactive allyl groups which, on weathering, react with each other under the influence of oxygen and light to bring about cross-linking. Consequently, 1,2-PBD hardens under weathering, in marked contrast to other biologically degradable polymers having photosensitizers. Thus, the photodegradation which begins simultaneously with sun exposure gives only coarse pieces and not a fine powder. These pieces do not injure plant growth, but instead work as a soil conditioner.[4] The products of incineration of 1,2-PBD are as expected, nontoxic.

References

1. Y. Takeuchi, A. Sekimoto and M. Abe, *ACS Div. Org. Coatings Plast. Chem. Pap.*, **34**(1), 122 (1974); *ibid.*, in R. D. Deanin, ed., *New industrial polymers, ACS Symp. Sec.*, **4**, 15 and 26 (1974).
2. Japan Pat. 564705; Japan Synth. Rubber Co.;
3. Japan Pat. 574706; Japan Synth. Rubber Co.;
4. *Japan Synthetic Rubber News*, **10**(1), 1 (1972).
5. Y. Takeuchi, *Japan Plastics Age*, **10**(7), 12 (1972).

2.9 POLY(PHENYLENE)

2.9.1 Structure and Synthesis

Hercules Inc. introduced in 1974 a family of thermosetting branched oligophenylenes[1-3] with the schematic structure[4]

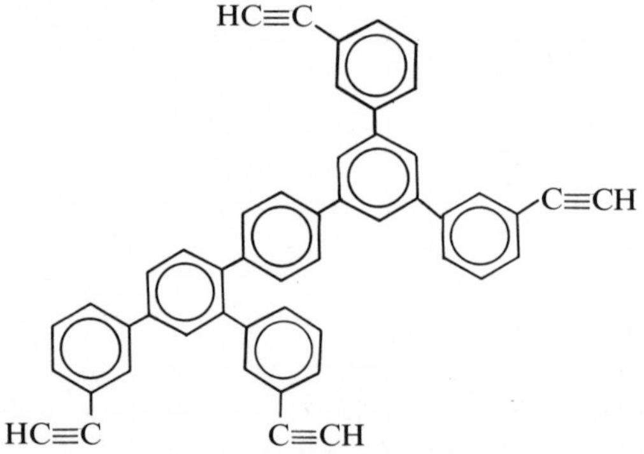

The soluble oligomers are marketed under the tradename Hercules H-resins®.

2.9.2 Properties and Application

The prepolymers are soluble in aromatic hydrocarbons, chlorinated hydrocarbons, ketones and cyclic ethers, usually in excess of 50 wt.-%. They are insoluble in water, aliphatic hydrocarbons, and aliphatic alcohols.

The prepolymers can be processed by a variety of fabrication techniques such as compression molding, injection molding, fiber extrusion, cold pressing, and power coating. Curing can be achieved by a number of catalysts, such as $TiCl_4/(C_2H_5)_2AlCl$ or nickel(II)acetylacetonate.[6] The fully cured polymers are hard, nonporous, and resistant to hot solvents, hot acids and bases, high temperature steam or brine, and molten salts. They become brittle in molten alkali metals. The cured polymers can be used continuously in air at temperatures of 200–300°C, and in oxygen-free environments at 400°C. Slow carbonization starts at 500–600°C.

Heating of the H-resins causes cross-linking, probably by cyclotrimerization of the acetylene endgroups to benzene rings. Form stability may be reached after curing at 160°C, but a post cure at 230–300°C is necessary for improved solvent resistance and mechanical properties.

H-resin moldings showed flexural strengths of 360–700 N/mm^2 and flexural moduli of 5000–8600 N/mm^2 at 23°C. These values improve on addition of fillers like talc, silica, or asbestos and decrease with increasing temperature.[3] The density of the cured, unfilled polymer is 1.145 g/cm^3, its limiting oxygen index 55 %, its volume resistivity 1017 ohm cm, and its dissipation factor 0.002 (at 60 Hz at 23 °C).[2]

Unmodified H-resins have low elongation and low adhesion to polar substrates. Special formulations have thus been developed for coatings. Such a coating (H-112) showed a Sward rocker hardness of 80 %, an elongation of 3 %, a linear thermal expansion of 2.9×10^{-5} K^{-1}, and a tensile bond strength of 3.2 N/mm^2.[4]

Adhesion of H-resins to metal surfaces can be improved by using polymer blends and polymer fillers such as epoxies, phenolics, polysulfones, or polyimides. Only polyimides can be used, however, as primers at temperatures above 250 °C.

H-resins are useful as corrosion resistant coatings, e.g. for pipes and processing equipment for hot sour wells, geothermal wells, high temperature sodium and lithium batteries, high temperature polymer and metal processing equipment, nonflammable coatings for electrical devices, lubricant coatings for metals, etc.[1-6]

References

1. L. C. Cessna, Jr. and H. Jabloner, *J. Elastomers Plastics*, **6**, 103 (1974).
2. H. Jabloner, L. C. Cessna and R. H. Mayer, *ACS Coatings and Plastics Preprints*, **34**(1), 198 (1974).
3. T. M. Bednarski, J. H. Del Nero, R. H. Mayer and J. A. Hagan, *SPE Ann. Techn. Papers*, **21**, 90 (1975).
4. J. French, *ACS Coatings and Plastics Preprints*, **35**(2), 72 (1975).
5. L. C. Cessna and H. Jabloner, *Rev. Gen. Caout, Plast.*, **52**(3), 103 (1975).
6. H. Jabloner and L. C. Cessna, Jr., *ACS Polymer Preprints*, **17**, 169 (1976).

2.10 POLY(OLEFIN-CO-VINYL ALCOHOL)

Nippon Synthetic Chemical Industry Co. introduced in 1974 a new engineering plastic under the tradename GL resin which is said to be an olefin-vinyl alcohol type copolymer.[1,2] Standard grades include a nonreinforced type (GL-N) and two glass-fiber reinforced types with 25 and 40% glass fiber, respectively (GL-G25 and GL-40).

GL resins are hydrophillic (see Table 2-10) and are thus not affected by aromatics, ketones, ethers, and halogenated hydrocarbons and the higher alcohols. They are resistant to fluorinated hydrocarbons, but methanol attacks them. The polymers absorb weak acids, weak alkalis, and dilute salt solutions, but remain largely unaffected by them.

The tensile and flexural strengths are higher than those of polycarbonate,

TABLE 2-10
Properties of GL resins.[1]

	Physical unit	Type		
		GL-N	GL-G 25	GL-G 40
Glass fiber content	%	0	25	40
Density	g/cm^3	1.215	1.41	1.52
Water content	%	0.3	0.25	0.20
Melting temperature	°C	186	186	186
Heat distortion temperature (30 min with no load)	°C	185	186	186
Thermal expansion coefficient	K^{-1}	4.1×10^{-5}	1.6×10^{-5}	0.94×10^{-5}
Tensile strength	N/mm^2	91	167	212
Elastic modulus	N/mm^2	44100	9800	13000
Elongation	%	40	8.5	8.2
Flexural strength	N/mm^2	130	230	300
Flexural modulus	N/mm^2	3900	10300	12900
Impact strength (without notch)	N/cm	53	68	109
Impact strength (with notch)	N/cm	392	392	392
Compression strength	N/mm^2	127	168	200
Flexural fatigue strength (10^7 times)	N/cm^2	35	>49	>49
Surface hardness (Rockwell)	—	R 120 M 100	R 120 M 102	R 120 M 104
Dielectric constant (10^6 Hz)	—	6.2	6.0	
Volume resistivity (1 kV)	Ω cm	2.9×10^{14}	5.7×10^{15}	
Surface resistivity	Ω	6.6×10^{12}	1.8×10^{15}	
Dielectric strength	kV/mm	30	34	
Dielectric loss tangent (10^6 Hz)	—	0.095	0.096	
Arc resistance	s	125	125	
Water absorption (20°C, 65% RH)	%	0.3	0.25	0.20
(20°C, sat. water)	%	7.5	6.2	4.9

nylon 6 and polyacetals. Creep resistance is also in general better. As usual, all mechanical properties can be improved by glass fiber addition (Table 2-13).

The hydrophilicity of the GL resins causes low dielectric properties on the one hand, but good permanent antistatic behavior on the other. Its anti-tracking behavior is equivalent to that of polyacetal and ABS and far superior to nylon, polycarbonate and phenolic resin.

References

1. H. Kawaguchi and T. Iwanami, *Jap. Plastics*, **8**(6), 6 (1974).
2. H. Kawaguchi and T. Iwanami, *Jap. Plastics*, **9**(1), 11 (1975).

2.11 NEW THERMOPLASTIC ELASTOMERS

2.11.1 Structure

Thermoplastic elastomers can be processed like thermoplastics above a certain temperature but possess rubber-like properties below this temperature. They achieve these characteristics through physical cross-links, i.e. domains of "hard" segments which melt or soften at a temperature above the service temperature but below the processing temperature. These domains are embedded in a matrix composed of "soft" segments. Thermoplastic elastomers thus need not be vulcanized chemically.

These properties can be obtained by several types of molecular architecture. Segmented polymers consist of many alternating hard and soft segments. Block copolymers are made up of long, hard segments connected to a soft segment. Graft copolymers consist of a backbone chain on which long side chains are grafted. Each molecular architecture can in turn utilize many different chemical building blocks (see Table 2-11).

The oldest thermoplastic elastomers are certain types of polyurethanes which are now manufactured by many companies. Similar in molecular architecture (but not in chemical composition) are DuPont's new Hytrel® polymers (see Chapter 6.6).

The styrene–butadiene–styrene block copolymers were introduced in 1965. They set the present standard of economics for the thermoplastic elastomers but lacked a good weatherability. Shell's second generation Kraton G® types overcome this deficiency: they have as center block a saturated polyolefin segment instead of a polybutadiene block. The Kraton G 7000® types are compounds of Kraton G with plasticizers, various resins and sometimes also inert inorganic fillers. Shell's Elexar® elastomers are apparently specially formulated Kraton G elastomers for the cable industry.[1]

The Solprene polymers of Phillips are radial block copolymers[2] with up to four branches.[3] They are manufactured by the reaction of styrene/diene diblock copolymers with, for example, silicon tetrachloride

$$R(\text{—CH—}\underset{\underset{C_6H_5}{|}}{\text{CH}}\text{—})_m(\text{diene})_n\text{Li} + \text{SiCl}_4 \rightarrow R(\text{—CH}_2\text{—}\underset{\underset{C_6H_5}{|}}{\text{CH}}\text{—})_m(\text{diene})_n]_4\text{Si} + 4\text{LiCl} \qquad (2\text{-}2)$$

Allied's ET polymer has the soft segments (butyl rubber) grafted on the hard segments (polyethylene)[4] whereas it is just the opposite for Hitachi's AAS polymers: hard segments (styrene/acrylonitrile) grafted on a soft backbone of saturated acrylic rubber.[5]

DuPont's Somel® is based on ethylene/propylene chemistry. The composi-

TABLE 2-11
Classes of thermoplastic elastomers.

Molecular architecture	Chemical composition	Tradenames	Companies
Segmented polymers	various hard and soft segments with urethane linkages	many	many
	soft (multibutyleneoxy)terephthalate segments and hard butyleneterephthalate segments (see Section 6.6)	Hytrel	DuPont
Block copolymers	Styrene-butadiene-styrene triblock	Kraton	Shell
	hydrogenated sty–bu–sty triblock	Kraton G	Shell
	star-like multiblocks with up to 4 butadiene center blocks and styrene end blocks	Solprene	Phillips Chemical
	?	TPR	Uniroyal
	?	TPN	DuPont
Graft copolymers	Segmented graft copolymer of butyl rubber on polyethylene	ET Polymers	Allied
	Styrene/acrylonitrile grafted on saturated acrylic rubber	AAS Polymers	Hitachi
unknown	Ethylene/propylene	Somel	DuPont
	?	Telcar	Goodrich

tion and molecular architecture of Uniroyal's TPR®, DuPont's TPN, and Goodrich's Telcar® have not been disclosed.

2.11.2 Properties and Application

Properties of thermoplastic elastomers vary widely, depending on chemical composition and molecular architecture (Table 2-12). They all do not need to be vulcanized. The absence of irreversible chemical cross-links allows scraps and trimmings to be reused. Fabrication can be carried out by injection molding or extrusion. Processing via solution leads in general to less favorable properties.

Thermoplastic elastomers are recommended for application in widely varying areas, such as adhesives, coatings, hoses, flexible containers, elastic bands, rubber toys, nonskid floor coverings, appliance gaskets, shoe soles, conveyor belts, and erasers. Solprene® is also used for rotationally molded industrial truck tires and other hollow parts.

TABLE 2-12
Physical properties of new thermoplastic elastomers.

Property	Physical unit	Property values for Kraton G Ref. 1	Telcar 100 Ref. 7	Somel 402T Ref. 6	AAS polymer Ref. 5	Solprene 406 Ref. 8
Density	g/cm^3	0.90	0.89	0.89		0.95
Heat distortion temp.	°C				85	
Softening temp. (Vicat)	°C		59			
Glass transition temperature (higher)	°C					102
Brittleness temperature	°C		< -80	< -76		
Glass transition temperature (lower)	°C					−92
Tensile strength at yield[a]	N/mm^2	9.2–14.8	13.0		37–42	
Tensile strength at break[a]	N/mm^2		13.0	16–31		27
Ultimate elongation[a]	%	700–825	290	160–595		700
Tensile modulus[a] at 100%	N/mm^2		12.8		1950–2250	
at 300%	N/mm^2					4.3
Flexural modulus	N/mm^2		165	385		
Stiffness	N/mm^2		242	262		
Compression set (22 h)		130–190	45			
Set at break		55–110				
Tear	N/mm	23–33		11.5		
Falling ball rebound	%	50–55				
Flexural strength	N/mm^2	6.5–9.9				
Hardness (Shore A)		50–60	93			90
(Shore D)				50		
(Rockwell R)					93–105	
Water absorption (24 h)	%			0.03		

[a]Property values depend on speed and method used.

References

1. Product information literature, Shell Chemical Co., 1974.
2. A. Thorsrud, *Modern Plastics*, **52**(10A), 96 (1975) (=Modern Plastics Encyclopedia 1975/76).
3. Anonym., *Chem. Week* (June 11, 1975), p. 35.
4. Anonym., *Polymer News*, **2**(9–10), 28 (1976).
5. H. Kohkame, M. Goto, Y. Muroi, *SPE Ann. Techn. Papers*, **20**, 449 (1974).
6. Product information literature, DuPont Co.
7. Product information literature, B. F. Goodrich Chemical Co.
8. J. R. Haws, in R. D. Deanin, ed., *New industrial polymers, ACS Symp. Ser.*, **4**, 1 (1974).

3 Unsaturated Carbon Chains

3.1 BROMINATED BUTYL RUBBER

3.1.1 Structure and Synthesis

Butyl rubber is a copolymer of isobutylene with a small amount of isoprene, that is, it contains groups derived from isobutylene (I) and isoprene (II). Brominated butyl rubber contains an additional 1.8–2.4 wt.-% bromine in the form of the groups III. It is produced by the Polymer Corporation, Ltd., Sarnia, Canada and is marketed under the tradename Bromobutyl®.

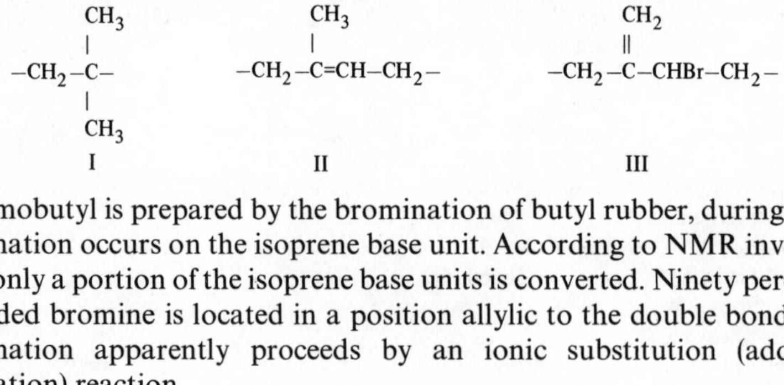

$$\underset{\text{I}}{-CH_2-\overset{\displaystyle CH_3}{\overset{|}{\underset{\underset{\displaystyle CH_3}{|}}{C}}}-} \qquad \underset{\text{II}}{-CH_2-\overset{\displaystyle CH_3}{\overset{|}{C}}=CH-CH_2-} \qquad \underset{\text{III}}{-CH_2-\overset{\displaystyle CH_2}{\overset{\|}{C}}-CHBr-CH_2-}$$

Bromobutyl is prepared by the bromination of butyl rubber, during which bromination occurs on the isoprene base unit. According to NMR investigations, only a portion of the isoprene base units is converted. Ninety percent of the added bromine is located in a position allylic to the double bonds. The bromination apparently proceeds by an ionic substitution (addition–elimination) reaction.

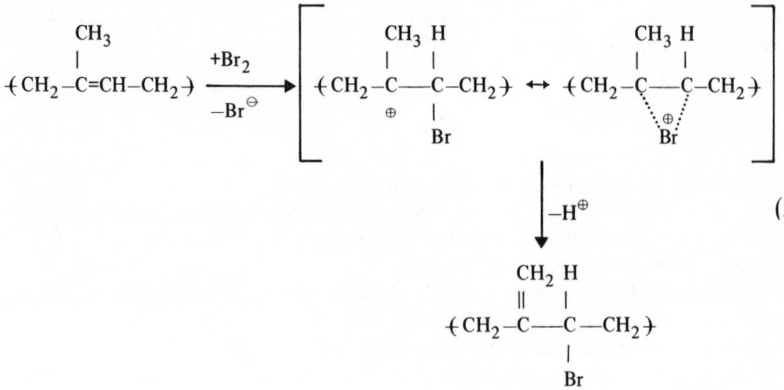

$$\left(CH_2-\overset{\displaystyle CH_3}{\overset{|}{C}}=CH-CH_2\right) \xrightarrow[-Br^{\ominus}]{+Br_2} \left[\left(CH_2-\overset{\displaystyle CH_3}{\overset{|}{\underset{\oplus}{C}}}-\overset{\displaystyle H}{\overset{|}{\underset{\underset{\displaystyle Br}{|}}{C}}}-CH_2\right) \leftrightarrow \left(CH_2-\overset{\displaystyle CH_3}{\overset{|}{C}}-\overset{\displaystyle H}{\overset{|}{C}}-CH_2\right)\text{(bridged }Br^{\oplus}\text{)}\right] \xrightarrow{-H^{\oplus}} \left(CH_2-\overset{\displaystyle CH_2}{\overset{\|}{C}}-\overset{\displaystyle H}{\overset{|}{\underset{\underset{\displaystyle Br}{|}}{C}}}-CH_2\right) \qquad (3\text{-}1)$$

3.1.2 Properties and Application

Butyl rubber possesses outstanding aging resistance and minimal permeability to air. It vulcanizes quite slowly as a result of the relatively few double bonds, and it shows rather poor adhesion. Both properties are improved by the addition of bromine or chlorine into the butyl rubber. In an older process butyl rubber was brominated batchwise in Banbury mixers with brominating agents. This product, however, was of low consistency and stability. Because of this and its high price, it was soon replaced by chlorinated butyl rubber. Chlorinated butyl rubbers are produced by the continuous chlorination in solution of butyl rubber. However, brominated butyl rubbers have higher vulcanization rates and better adhesion properties than the chlorinated products. The new brominated butyl rubbers are prepared by a continuous process and, therefore, may have a homogeneous distribution of bromine atoms and subsequently a better stability than the older bromobutyl rubbers. In fact, the viscosity of raw bromobutyl rubber remains unchanged over 28 months.[1]

Brominated butyl rubber is homogeneous and free of gels. It vulcanizes faster than normal butyl rubber or its chlorinated derivative, since the allylic double bonds are more reactive than the centrally located double bonds of the isoprene units, and also because the bromine participates in the vulcanization. For this reason, in contrast to normal butyl rubber, bromobutyl rubber can be co-vulcanized with natural rubber or butadiene/styrene rubber. In mixtures of normal butyl rubber with, for instance, natural rubber, the base units of the natural rubber react much faster with the vulcanizing agent than the isoprene units of the butyl rubber. Consequently, the butyl rubber remains undervulcanized. Blends with saturated elastomers such as EPDM are also possible. The vulcanization can be carried out with sulfur and sulfur compounds, zinc oxide, lead oxide as well as with reactive phenolic resins.

The products are employed similarly to normal rubber, that is, primarily for tire inner tubes. Use as tire walls is also advantageous, since the brominated butyl rubber bonds well to most carcasses produced from natural rubber, in contrast to butyl rubber. A number of other use areas are based on the somewhat improved heat stability, for instance as sealants and conveyor belts.

References

1. J. Walker, R. H. Jones and G. Feniak, Rubber Div. Meeting, Canadian Institute of Chemistry, Toronto, May 26, 1972.
2. Product information literature, Polymer Corporation, Ltd., Sarnia, Canada.
3. W. J. van der Veen, *Plastica*, **28**, 236 (1975).

TABLE 3-1
Properties of typical vulcanizates of Bromobutyl, butyl rubber and natural rubber.[2]

Property	Physical unit	Values measured for Bromobutyl	Butyl rubber	Natural rubber
Tensile stress	N/mm^2	7.8	11.4	21.4
Ultimate elongation	%	570	670	560
Hardness (Shore A)		45	53	52
After 22 h aging at 177°C in air				
Tensile stress	N/mm^2	4.1	0.3	2.4
Ultimate elongation	%	300 (dry)	550 (sticks)	220 (sticks)
Hardness (Shore A)		63	60	44
After 70 h aging at 177°C in air				
Tensile stress	N/mm^2	1.9	Decomposition	Decomposition
Ultimate elongation	%	230 (dry)		
Hardness (Shore A)		68		

3.2 1, 2-1, 4-POLY(BUTADIENES)

3.2.1 Structure and Synthesis

In the polymerization of butadiene with lithium or lithium alkyls, variation of solvent and temperature and/or the addition of Lewis bases affords a whole series of poly(butadienes) with varying amounts of 1,2-vinyl groups in the polymer (Table 3-2). The development of these types of 1,2-1,4-poly-

TABLE 3-2
Chemical structure of poly(butadiene) prepared from various lithium initiators (according to data from Ref. 1).

Initiator	Solvent	Structure in % *cis*-1,4	*trans*-1,4	1,2
LiC_2H_5	Hexane	43	50	7
LiC_2H_5	Toluene	44	47	9
Li	Hexane	35	55	10
LiC_2H_5	Benzene/$(C_2H_5)_3N$	23	40	37
Li	$(C_2H_5)_2O$	16	20	64
LiC_2H_5	Benzene/tetrahydrofuran	13	13	74
LiC_2H_5	Heptane/tetrahydrofuran	4	4	92
LiC_2H_5	Tetrahydrofuran	0	9	91

(butadienes) is of interest[2–3] because poly(butadienes) with vinyl group contents between 35 and 55% have abrasion, elasticity, friction and traction properties similar to the blends of styrene/butadiene rubbers (SBR) with *cis*-1,4-poly(butadiene) (BR) (Figure 3-1).

However, blends of these two rubbers have the disadvantage that the mechanical properties depend on the dispersion of the two phases, which is inherently impossible with 1,2-1,4-poly(butadiene). In addition, 1,2-1,4-poly(butadiene) should be less dependent on the raw material situation than SBR, whose production fluctuates with the availability of styrene. From time to time in recent years, styrene has been in short supply, because the benzene required for its production has been in demand to improve the anti-knock properties of lead-free gasoline.

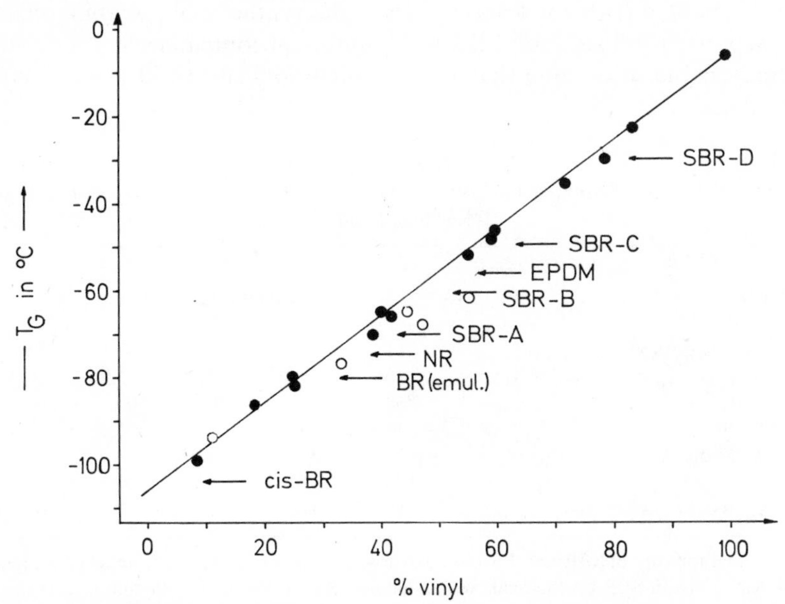

Figure 3-1 Dependence of the glass temperature T_G of 1,2-1,4-poly(butadienes) on the vinyl-group (1,2-structure) content according to data from Ref. 2 (○) and Ref. 3 (●). Also plotted are the glass temperatures of other elastomers (see Ref. 1). The letters after the abbreviation SBR have the following meanings: A (19% styrene units, 8% vinyl units), B (25% styrene units, 8% vinyl units), C (25% styrene units, 25% vinyl units) and D (40% vinyl units).

The pilot plant run by the firm Chemische Werke Hüls operates under adiabatic polymerization conditions.[2] Heat of polymerization causes the temperature to rise to about 150°C which is favorable to the property traits of the elastomer. The technically desired long-chain branching is obtained at the high temperatures but not at temperatures under 80°C. Long-chain branching influences the elasticity.

The high polymerization temperature also drastically shortens the polymerization time to about 3 min. In addition, broader molecular weight distribution is obtained by the pulsating addition of alkyllithium compound. The adiabatic reaction conditions lead to a nonuniform distribution or blocking of the vinyl groups along the polymer main chain.

3.2.2 Properties and Fabrication

As is apparent from Figure 3-1 and Table 3-3, 1,2-1,4-poly(butadienes) with vinyl group contents between about 33 and 53% show approximately the same glass temperatures and the same abrasion and traction as conventional mixtures of SBR and BR (Table 3-3). Alkyllithium initiators have been used on a commercial scale for several years in the synthesis of poly(butadienes), poly(isoprenes) and solution-SBR by a number of companies; consequently, it is reasonable to assume that 1,2-1,4-poly(butadienes) will succeed commercially, also.

TABLE 3-3
Properties of 1,2-1,4:poly(butadienes) as a function of vinyl-group content (according to data from Refs. 1 and 3).

Property	Physical unit	Values measured for the samples				
Content of 1,2-groups	%	11	33	45	47	55
Tensile strength	N/mm^2	16.2	16.4	17.9	15.7	16.0
Ultimate elongation	%	560	490	460	590	600
Modulus (300%)	N/mm^2	6.5	8.4	8.6	5.8	5.8
Hardness (Shore A)	—	57	61	59	58	57
Abrasion[a]	—	130	100	88	91	77
Skid resistance[a]	—	85	96	102	100	103

[a]Relative to a mixture of SBR and BR (60/40) whose abrasion and skid resistance were each set at 100. An emulsion-SBR on this scale would have a relative abrasion of 90 and a relative skid resistance of 107.

1,2-1,4-poly(butadienes) can be produced in powder form[4] The processing of elastomers is known to be extraordinarily labor and energy intensive, so that for a long time the development of a powdered or a liquid elastomer has been sought. Powdered or liquid rubbers are considerably more simple to fabricate, especially with regard to the expensive blending of additives. In fact, the number of processing steps is much smaller for liquid elastomers (compare Section 3.3) than for powdered elastomers. However, a liquid, all-purpose elastomer has yet to be developed. Powdered elastomers may be seen, therefore, as the next step along the way from conventional bale rubber to liquid rubber.

Powdered rubber, because of the cold-flow and the self-adhesion of the rubber, cannot be produced in a form suitable for storage either by spray drying, freeze drying, flash evaporation, milling, microencapsulation or by introducing powdery cover layers. The coprecipitation of elastomers and fillers into master batches with particle sizes of 100–1500 μm appears promising. In addition to the solution poly(butadiene), the important EPM and EPDM rubbers as well as the SBR latices can be treated according to this method, also. Table 3-4 shows for a butadiene rubber with 35% 1,2-groups that the mechanical properties of the end product are practically independent of whether the starting material was (a) baled rubber or (b) and (c) powdered rubber. (a) and (b) were processed conventionally on calenders; (c) in contrast, was processed by direct feed of the vulcanization-ready powdered mixtures into extruders with special screws.

TABLE 3-4

Mechanical properties of vulcanizates of 1,2-1,4-poly(butadiene) containing 35% vinyl groups as a function of its fabrication. (a) bale-form rubber processed with calenders, (b) powdered rubber processed with calenders, (c) powdered rubber processed with special extruders (according to Ref. 4).

Property	Physical	Values measured for fabrication according to method (a)	(b)	(c)
Tensile strength	N/mm^2	14.9	14.5	13.2
Ultimate elongation	%	530	510	515
Modulus (300%)	N/mm^2	7.4	7.3	6.5
Shore hardness		61	61	64
Falling ball rebound	%	32	30	33
Abrasion index (12,000 km)		100	103	100

References

1. H. E. Adams, R. L. Bebb, L. E. Rorman and L. B. Wakefield, *Rubber Chem. Technol.* **45**, 1252 (1972).
2. K.-H. Nordsiek and N. Sommer, *Der Lichtbogen*, **23**(3) (No. 174), 8 (1974), (journal of the firm Chemische Werke Hüls).
3. A. E. Oberster, T. C. Bouton and J. K. Valaitis, *Angew. Makromol. Chem.*, **29/30**, 291 (1973).
4. G. Berg and K.-H. Nordsiek, Der Lichtbogen **23**(3) (No. 174), 12 (1974), (journal of the firm Chemische Werke Hüls).

3.3 LIQUID RUBBERS

Conventional rubber fabrication is very expensive as a result of the very high viscosity of the polymers: the blending in of the vulcanization catalysts, activators, aging stabilizers, pigments, fillers, plasticizers and so forth must be carried out on mixing rollers and the vulcanization mostly in presses. Liquid rubbers, on the other hand, offer considerable advantages.[1-5] Although fluid elastomers such as the silicones, the polyethers and the polyesters have been known for some time (compare Table 3-5), only recently has a fluid rubber based on butadiene been brought onto the market (Table 3-6). The total production of fluid rubbers, with the exclusion of that used for rocket propellants, may currently amount to about 30,000 t/a.[1,4]

TABLE 3-5
Commercial liquid rubbers with varying types of base units (according to compilations in Ref. 2).

Monomer unit in the main chain	Cross-linkable group	Manufacturer	Tradename	Price[a] \$/kg
Chloroprene	Cl	DuPont	Neoprene FB, FC	1.43
Isobutylene	$>C=C<$	Enjay	Butyl LM 430	1.54
Isoprene (nat.)	$>C=C<$	Hardman	DPR	1.34
(synth.)	$>C=C<$	Hardman	Isolene D	1.30
Ester	NCO	many	—	1.92
Ether	NCO	many	—	2.14
Sulfide	SH	Thiokol	LP	1.83
Siloxane	$>C=C<$	Dow Corning, UCC, GE, Stauffer, Wacker, Mobay, and others	—	3.63

[a]As of 1973.

The simplest types of liquid rubbers are made up of the degradation products of, for instance, poly(isoprene). Their cross-linking takes place, therefore, through the residual carbon–carbon double bonds. Emphasis lies, however, on the development of liquid rubbers with reactive endgroups. In the anionic polymerization of butadiene, for instance, using bifunctional initiators polymer chains are formed having two anionic growth centers, which may be converted on termination with CO_2 into carboxyl endgroups:

$$\sim\sim CH_2-CH=CH-CH_2^{\ominus} + CO_2 \longrightarrow \sim\sim CH_2-CH=CH-CH_2-COO^{\ominus} \quad (3\text{-}2)$$

TABLE 3-6
Commercial liquid rubbers based on butadiene (according to compilations in Ref. 2).

Monomer unit in the main chain	Cross-linkable group	Manufacturer	Tradename	Price[a] $/kg
Butadiene	$>C=C<$	Lithium Corp.	Litherne A, P, Q	0.77
		Richardson	Ricon 150	1.54
	OH	Phillips	Butarez HTS	9.36
		Arco	Poly bd R-45	1.10
		Hystl Dev.	Hystl G	1.67
	COOH	Hystl Dev.	Hystl C	1.67
		Phillips	Butarez CTL II	9.36
		Goodrich	Hycar CTB	3.30
		Thiokol	HC-434	7.01
	BR	Polymer Corp.	RTV Liquid Rubber	2.20
Butadiene/acrylonitrile	$>C=C<$	Goodrich	Hycar 1312	1.12
	OH	Arco	Poly bd CN-15	1.23
	COOH	Goodrich	Hycar CTBN	2.75
	SH	Goodrich	Hycar MTBN	2.53
Butadiene/styrene	$>C=C<$	ASRC	Flosbrene 25	0.75
		Richardson	Ricon 100	1.54
	OH	Arco	Poly bd CS-15	1.10

[a]As of 1973.

Hydroxyl endgroups can be introduced by termination with ethylene oxide, for example. In every case a relatively high initiator concentration is used in order to obtain as nearly as possible the desired low molecular weight (between 3000–10,000 g/mol) which is necessary for the liquid rubbers. Vulcanization consists then of reaction of these reactive endgroups with polyfunctional compounds, for example, for liquid rubbers with hydroxyl endgroups, reaction with polyisocyanates.

There has been an extraordinary variety of cross-linking reactions known and proposed (compare, for instance, the compilation in Ref. 2). In general, the concentration of the cross-linker is relatively high. In addition, many cross-linking systems demand strict adherence to stoichiometric requirements. Several cross-linking systems work at room temperature, others require higher temperatures.

Liquid rubbers are already employed in a variety of applications (see, for instance, Ref. 2). The sought-after goal, the mass production of tires, is not yet realized, since the fluid rubbers are more expensive than normal rubber; they also leave something to be desired in their abrasion properties (see Table 3-7). However, they are already used in the retreading of tires.

TABLE 3-7
Physical properties of mixtures for tire treads (according to Ref. 3).

Property	Physical unit	Value measured for normal SBR	liquid SBR with COOH	liquid SBR with OH
Tensile strength	N/mm^2	24	15	15.5
Ultimate elongation	%	540	340	270
Modulus (200%)	N/mm^2	5.2	5.3	7.2
Hardness (IRHD)	—	61	63	85
Falling ball rebound	%	45	41	50
Abrasion (Akron-Test)	$cm^3/1000$ U	0.21	1.21	0.43
Temperature increase (K in 25 min)	K	49	74	100

References

1. P. Kitzmantel, *Allgem. Prakt. Chem.*, **24**(5-6), 93–96 (1973).
2. E. A. Sheard, *Chem. Technol.*, **3**, 298 (1973).
3. J. P. Berry and S. H. Morrell, *Polymer* [*London*], **15**, 521 (1974).
4. R. Koncos, *Rubber Chem. Technol.*, **46**, 590 (1973).
5. H. Singer, *Plast. Kautsch.*, **22**(5), 404 (1975).

3.4 POLY(PENTENAMER)

3.4.1 Structure and Synthesis

The polymers which result from the ring-opening polymerization of cyclopentene are designated as poly(pentenamers):

$$\text{cyclopentene} \longrightarrow \left(CH{=}CH{-}CH_2CH_2CH_2 \right) \qquad (3\text{-}3)$$

The name poly(pentenamer) goes back to an IUPAC nomenclature which has subsequently been abandoned. According to present-day IUPAC nomenclature, the polymer would be designated as poly(1-pentenylene). Many firms are involved with poly(pentenamer), in particular Bayer, Goodyear, Montedison and Phillips Petroleum. In 1972 Bayer announced the building of a 40,000 t/a facility, which was to be finished in 1974; completion has, however, apparently been postponed.

Cyclopentadiene results from the cracking of petroleum naptha and heavy petroleum fractions. It is separated from the C_5-cut by extractive distillation, during which approximately equal amounts of isoprene on the one side and cyclopentene, cyclopentadiene and dicyclopentadiene on the other side are collected. Cyclopentadiene can be dimerized to dicyclopentadiene

$$2 \rightleftarrows \qquad (3\text{-}4)$$

which, in turn, can be split and hydrogenated to give cyclopentene.

The polymerization of cyclopentene is carried out in solution with R_3Al WCl_6/C_2H_5OH to give *trans*-poly(pentenamer). The *cis*-poly(pentenamer) prepared, for example, from the catalyst $R_2AlCl/MoCl_5$ is of no commercial importance. The polymerization represents a special case of olefin metathesis and accordingly is governed by entropy. For this reason there is practically no heat generated during the polymerization. The majority of the product is acyclic, although large rings also occur. The synthesis and properties of poly(pentenamers) have been extensively covered in several review articles.[1-3]

3.4.2 Properties and Application

The *trans*-poly(pentenamer) is an all-purpose rubber, which resembles partly *cis*-poly(butadiene) and partly natural rubber. Like natural rubber it shows high tensile stress at maximum load (Figure 3-2), good adhesiveness, and crystallization on elongation. The latter has a reinforcing effect, that is with increasing deformation it causes an increase in the tensile stress at maximum load. Styrene/butadiene rubber and *cis*-1,4-poly(butadiene) do not show this effect (Figure 3-2).

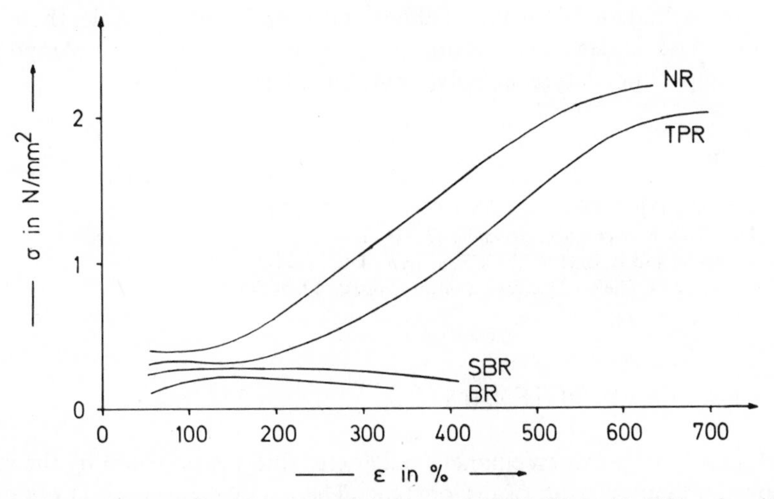

Figure 3-2 Crude strength σ of a *trans*-poly(pentenamer) TPR filled with 50% carbon-black as a function of deformation ε in comparison to natural rubber NR, butadiene/styrene-rubber SBR and *cis*-1,4-poly(butadiene) BR (according to Ref. 4).

Furthermore, *trans*-poly(pentenamer), like *cis*-1,4-poly(butadiene) can be filled with great amounts of oil and carbon black. It shows an excellent resistance to thermal and mechanical degradation and a low abrasion. Vulcanization may be carried out similarly to that of natural rubber, *cis*-1,4-poly-(butadiene) or styrene/butadiene rubber (see Ref. 4). After an initial period, the properties of the vulcanized product are largely independent of vulcanization time (Table 3-8).

TABLE 3-8

Influence of vulcanization time on several properties of *trans*-poly(pentenamer)-vulcanizate (according to Ref. 2). Mixture out of 100 parts polymer, 1.5 parts PBNA, 2 parts stearic acid, 5 parts zinc oxide, 30 parts Circosol-oil 4240, 50 parts ISAF-carbon black, 0.5 parts TMTD, 1.5 parts Sulfasan-R.; vulcanization at 140°C.

Property	Physical unit	Values measured at vulcanization times of				
		20 min	40 min	80 min	120 min	240 min
Tensile strength	N/mm^2	7.7	17.9	18.3	17.5	15.5
Ultimate elongation	%	540	360	360	360	340
Modulus (200%)	N/mm^2	1.6	6.1	6.5	6.2	6.2
(300%)	N/mm^2	3.1	13.2	13.9	12.8	13.0
Hardness (IRHD)		39	65	64	64	63
Tear	N/cm	210	440	440	440	390

In addition, *trans*-poly(pentenamer) can be mixed in any desired ratio with every other known all-purpose rubber and is eminently suitable, therefore, for tires. Also under consideration is its use as the elastomeric component of high impact poly(styrene), poly(vinyl chloride) and ABS.

References

1. A. J. Amass, *Brit. Polymer J.*, **4**, 327 (1972).
2. G. Dall'Asta, *Rubber Chem. Technol.*, **47**, 511 (1974).
3. N. Calderon and R. L. Hinrichs, *Chem. Tech.*, **4**, 627 (1974).
4. F. Haas and D. Theisen, *Kautsch. Gummi Kunstst.*, **23**, 502 (1970).

3.5 POLY(OCTENAMER)

Analogously to poly(pentenamer), poly(octenamer) is prepared by the ring-opening polymerization of cyclooctene. The cyclooctene itself is obtained from the hydrogenation of cyclooctadiene, which is the cyclodimerization product of butadiene:

$$2CH_2{=}CH{-}CH{=}CH_2 \rightarrow \qquad \xrightarrow{+H_2} \qquad \rightarrow \, +CH{=}CH{-}(CH_2)_6+ \qquad (3\text{-}5)$$

Commerical development appears to have concentrated on two varieties. Poly(octenamers) with *cis*-contents of 75–80% have been developed at Montedison,[1] while at Chemische Werke Hüls the development work has centered around polymers with *cis*-contents of 40–50%.[2] Although the *cis*-poly(octenamer) has about the same melting point as the *trans*-poly(pentenamer), it has a far higher rate of crystallization, which is lowered by reduction in the *cis*-content. Both types appear to have good green strength; both are considered for use as an all-purpose rubber.

The polymer with the higher *cis*-content is relatively difficult to process on open mills,[1] whereas the Hüls product is said to have excellent processability.[2,3] In contrast to a *cis*-poly(butadiene) of higher molecular weight, the Hüls polymer shows higher tensile strength, modulus and resilience (Table 3-9). This somewhat surprising discovery is attributed to the considerably reduced number of chain ends, for, according to analytical investigations about 40% of the *cis*-poly(octenamer) consists of macrocyclic rings.[2].

TABLE 3-9

Comparison of the properties of a vulcanized poly(octenamer) having a 45% *cis*-content and a viscosity average molecular weight of 65,000 g/mol with those of a vulcanized *cis*-poly(butadiene) having a 92% *cis*-content and a viscosity average molecular weight of 109,000 g/mol.[2] Mixture: 100 parts elastomer, 50 parts Corax 3, 10 parts Naftolen MV, 3 parts ZnO RS, 2 parts stearic acid and 1 part PBN, as well as 1.2 parts sulfur and 0.8 parts Vulcazit-CZ (poly(octenamer)) or 1.5 parts sulfur and 0.9 parts Vulcazit-CZ (poly(butadiene)). Vulcanization was carried out at 150°C for the times listed.

		Values measured for			
		poly(octenamer)		poly(butadiene)	
Property	Physical unit	50 min	100 min	30 min	90 min
Tensile strength	N/mm²	12.0	10·8	6.6	6.3
Ultimate elongation	%	334	282	358	410
Modulus (300%)	N/mm²	10.5	—	5.3	4.3
Permanent extension	%	7	5	11	13
Hardness (Shore A)		65	67	55	52
Falling ball rebound	%	42	43	30	28
Abrasion (DIN)	mm³	96		82	

References

1. G. Dall'Asta, *Rubber Chem. Technol.*, **47**, 511 (1974).
2. W. Holtrup, F.-W. Küpper, W. Meckstroth, H.-H. Meyer and K.-H. Nordsiek, *Der Lichtbogen*, **23**(174), 16 (1974), (journal Chemische Werke Hüls).
3. W. J. van der Veen, *Plastica*, **2** (11), 516 (1974).

3.6 POLY(BUTADIENE-CO-VINYLIDENE CHLORIDE)

The Dow Chemical Company recently offered, in experimental quantities, a range of copolymers of butadiene and vinylidene chloride as halogenated, self-cross-linking latices for the carpet industry. These latices (Experimental Latices XD-8601, XD-8602 and XD-8603) are intended to serve principally in the lamination of jute for tufted rugs. The solids content of the latices is about 48 % with a pH 7.8 and a surface tension of $(5.5–6.5) \times 10^{-4}$ N/cm. The latices can be loaded with fillers up to a proportion of 4:1. The tensile strength of the filled films from these latices is about 3.5 N/mm^2, and the ultimate elongation is about 10–15 %. Calcium carbonate/aluminum oxide trihydrate-filled films show limiting oxygen indices LOI of about 29–35, according to the proportion and mixing ratio of the filler.[1]

References

1. Product information literature, The Dow Chemical Co.

3.7 POLY(NORBORNENE)

Norbornene (bicyclo[2,2,1]heptene-2) is the Diels–Alder addition product of ethylene and cyclopentadiene. It polymerizes under ring-opening to poly(norborene) with molecular weights of two million:[1-4]

$$\text{norbornene} \xrightarrow{WCl_6/AlP_3/J_2} \left(\text{cyclopentane-1,3-diyl}{-}CH{=}CH \right) \qquad (3\text{-}6)$$

The double bond may be in a *cis* or *trans* position. The polymer is manufactured by CdF Chimie (= a subsidiary of Charbonnages de France) under the tradename Norsorex® since 1975.

The stiff chain leads to thermoplastic behavior: melting points are reported as 170–190°C and glass transition temperatures as 45–47°C[1] or 35°C.[3] The polymer can absorb mineral oil up to four times its weight whereupon it shows rubber-like behavior with glass transition temperatures of −45 to −60°C. Vulcanization can be carried out with classic recipes, e.g. with sulfur, ZnO and the conventional activators. A typical recipe[3-4] of 100 parts Norsorex, 200 parts aromatic oil, 200 part SAF black, 0.5 parts sulfur, 6 parts CBS (activator), 5 parts ZnO, and 2 parts stearic acid can be cured 10 min at 155°C. Such a cured elastomer exhibits the following properties: tensile strength of 14.5 N/mm^2, elastic moduli of 1.8 and

11.8 N/mm^2 at 100 and 300% extension, respectively, elongation at break 375%, hardness 65° on the Barkol scale, resilience 20% at 50°C, and low temperature stiffening at −12°C.

Norsorex is suggested for use in vibration and noise damping systems and conveyor belts.

References

1. German Patent 1961742 (Dec. 9, 1969/June 18, 1970); French priority of December 9, 1968; Societé Chimique de Charbonnages, France; Inv.: J. C. Muller.
2. C. Stein and A. Marbach, *Plast. Mod. Elastomères*, **26**(9), 78 (1974).
3. C. Stein and A. Marbach, *Rev. Gen Caout. Plast.*, **52**(1-2), 71 (1975).
4. W. Cooper, *Europ. Polymer J.*, **11**, 833 (1975).

4 Fluoropolymers

The rapidly expanding area of fluoropolymers has been frequently treated in past years in review articles; consequently, only the newest developments need be considered here.

Review Articles

O. Scherer, *Fluorokunststoffe*, *Fortschr. Chem. Forschg.*, **14**, 161 (1970).

L. A. Wall (Ed.), *Fluoropolymers* (High Polymers, Vol. XXV), Wiley-Interscience, New York, 1972.

R. G. Arnold, A. L. Barney and D. C. Thompson, Fluoroelastomers, *Rubber Chem. Technol.*, **46**, 619 (1973).

4.1 POLY(TETRAFLUOROETHYLENE-CO-ETHYLENE)

4.1.1 Structure and Synthesis

As is well known, poly(tetrafluoroethylene) has excellent properties, but it is difficult to process. In search of a more easily fabricated fluoropolymer, several companies developed copolymers of tetrafluoroethylene and ethylene. The copolymerization parameters for the radical polymerization are at $-30°C$

$$r_{C_2F_4} = 0.013 \pm 0.008,\ r_{C_2H_4} = 0.10 \pm 0.02$$

at $+65°C$,

$$r_{C_2F_4} = 0.045 \pm 0.007,\ r_{C_2H_4} = 0.14 \pm 0.02$$

Hence it follows that at low temperatures there is a tendency to alternating copolymerization, which is reduced at higher temperatures. The commercial products from DuPont (Tefzel®), Hoechst (Hostaflon ET®) and Montedison are largely alternating. DuPont reports a tetrafluoroethylene content of more than 75 wt.-%, that is of more than 1:1 = mol/mol.[2] Hoechst describes its product as an alternating copolymer with more than 90% alternation.[3,4] Montedison's copolymer has more than 55 mol-% tetrafluoroethylene.[1] The alternation was determined by infrared and X-ray measurements. 1:1 copolymers also show a maximum in melt temperature as a function of composition.[1] These alternating copolymers are viewed formally as head/head–tail/tail copolymers of vinylidene fluoride. Their degree of crystallization is estimated at 50–60 %. Some of the polymers are also offered reinforced with glass fibers.

4.1.2 Properties and Application

With the exception of reduced heat stability, the copolymers have practically all the properties of poly(tetrafluoroethylene). But they are easier to fabricate, have a lower density and show essentially lower creep.[3] The electrical proper-

TABLE 4-1
Properties of alternating copolymers of tetrafluoroethylene and ethylene.

Property	Physical unit	Values measured for			
		Hoechst Hostaflon ET Ref. 3	Montedison Ref. 1	DuPont Tefzel 200 Refs. 2, 5 and 6	DuPont Tefzel 200 + 25 wt.-% glass fibers Ref. 5
Density	g/cm^3	1.7–1.9		1.7	1.86
Glass temperature	°C		110		
Melting point	°C	270	275	271	
Brittleness temperature	°C		−80	< −101	
Heat distortion temperature (1.85 N/mm)	°C		46	71	210
Continuous service temperature	°C	170		150–180	
Tensile yield strength	N/mm^2			28.2	
Tensile strength	N/mm^2	30–55	50	45.5	85
Ultimate elongation	%		300	150	9
Modulus	N/mm^2		1400		
Flexural modulus	N/mm^2			1400	6700
Compression set	%	2			
Impact strength with notch	J/cm^2			no break	
Flex life	—		50,000–100,000	30,000	
Dielectric constant	—		2.4	2.6	
Dissipation factor	—		0.0002	0.0006	
Dielectric strength	kV/mm		25.9	> 80	
Volume resistivity	Ω cm			> 10^{16}	
Hardness (Shore D)	—		68	75	
(Rockwell)	—			50	
Friction constant	—			0.4	0.31
Flammability	—		flame retardant	flame retardant	flame retardant

ties are good. The coefficient of friction is low and the abrasion resistance is high. Like all fluoropolymers, the copolymers have moderate tensile strength (see Table 4-1). The stability of the copolymers to most organic and inorganic solvents is good.[1] They are stable also to amines, ketones, esters and conc. sulfuric acid, which attack poly(vinylidene fluoride).

Halogenated hydrocarbons attack poly(tetrafluoroethylene-co-ethylene) less than poly(chlorotrifluoroethylene). The copolymers are also radiation resistant.[2,6]

The copolymers may be processed by all the fabrication methods used for thermoplastics, for instance, injection molding, rotational molding, blow molding and extrusion. Montedison specifies fabrication temperatures of 300–330°C. Use of the polymers in the electrical industry (cable insulation, sockets, switches, batteries) as well as in the chemical industry (centrifugal pumps, laboratory apparatus) is foreseen.

References

1. C. Garbuglio, M. Modena, M. Valera and M. Raggazzini, *Europ. Polymer J.*, **10**, 91 (1974).
2. Product information literature, DuPont de Nemours.
3. H. Cherdron, *Chimia* [*Aarau*], **28**, 553 (1974).
4. H. Fitz, *Kunststoffe*, **62**, 647 (1972).
5. Anonym, *Kunststoffe*, **64**, 92 (1972).
6. J. C. Reed and J. R. Perkins, lecture, *21st International Wire and Cable Symposium*, Atlantic City, N.J., December 6, 1972.

4.2 POLY(TETRAFLUOROETHYLENE-CO-PERFLUORO(METHYL VINYL ETHER))

4.2.1 Structure

Under the designation ECD-006 are offered by DuPont research quantities of a terpolymer of tetrafluoroethylene (I), perfluoro(methyl vinyl ether) (II) and a small amount of a perfluorinated, cross-linkable monomer (III–VI):[1–6]

I	$CF_2{=}CF_2$	tetrafluoroethylene
II	$CF_2{=}CF(OCF_3)$	perfluoro(methyl vinyl ether)
III	$CF_2{=}CF{-}O{-}(CF_2)_4CN$	perfluoro(4-cyanobutyl vinyl ether)
IV	$CF_2{=}CF{-}O{-}(CF_2)_4COOCH_3$	"perfluoro"(4-carbomethoxybutyl vinyl ether)
V	$CF_2{=}CF{-}O{-}CF_2CF(CF_3){-}OC_6F_5$	perfluoro(2-phenoxypropyl vinyl ether)
VI	$CF_2{=}CF{-}O{-}(CF_2)_3{-}OC_6F_5$	perfluoro(3-phenoxypropyl vinyl ether)

The product ECD-006 contains about 60% tetrafluoroethylene; the structure of the cross-linked portion has not been revealed.[7]

4.2.2 Synthesis and Vulcanization

The terpolymer is produced through radical copolymerization either in fluorinated hydrocarbons or in aqueous emulsion.[8] Since the reactivities of the three radicals are approximately the same, it is relatively easy to obtain terpolymers with about 50–60% tetrafluoroethylene units. The commercial synthesis appears to make use of emulsion polymerization under pressure with persulfate or redox initiators and ammonium perfluorooctanoate. A transparent latex occurs in the terpolymerization in emulsion. This is coagulated either by freezing or by treatment with acids or salts.

The cross-linking required for the elastomeric properties proceeds in a different manner for each of the termonomers.[1] The nitrile (III) trimerizes to a triazine derivative under the influence of tetraphenyltin:

$$3 \sim\sim(CF_2)_4CN \rightarrow \text{triazine ring: } C_3N_3[(CF_2)_4\sim\sim]_3 \tag{4-1}$$

The ester (IV) is either transesterified with glycols or converted to amides with diamines:

$$2\sim\sim COOCH_3 + HO\text{–}R\text{–}OH \rightarrow \sim\sim COO\text{–}R\text{–}OOC\sim\sim + 2\,CH_3OH \tag{4-2}$$

$$2\sim\sim COOCH_3 + H_2N\text{–}R\text{–}NH_2 \rightarrow \sim\sim CONH\text{–}R\text{–}NH\text{–}OC\sim\sim + 2\,CH_3OH \tag{4-3}$$

The phenoxy compounds (V and VI) are vulcanized either with diamines + magnesium oxide or with the dipotassium salts of bisphenols, for example:

$$2\sim\sim CF(CF_3)\text{–}O\text{–}C_6F_5 + H_2N\text{–}R\text{–}NH_2 + MgO \rightarrow$$
$$\sim\sim CF(CF_3)\text{–}O\text{–}C_6F_4\text{–}NH\text{–}R\text{–}NH\text{–}C_6F_4\text{–}O\text{–}CF(CF_3)\sim\sim + MgF_2 + H_2O \tag{4-4}$$

$$2\sim\sim CF(CF_3)\text{–}O\text{–}C_6F_5 + KO\text{–}C_6H_4\text{–}C(CH_3)_2\text{–}C_6H_4\text{–}OK \rightarrow$$
$$\rightarrow \sim\sim CF(CF_3)\text{–}O\text{–}C_6H_4\text{–}C(CH_3)_2\text{–}C_6H_4\text{–}O\text{–}CF(CF_3)\sim\sim + 2\,KF \tag{4-5}$$

During fabrication foreign matter must be strictly excluded.

4.2.3 Properties and Application

The properties of the raw polymer are uninfluenced by hydrocarbons, alcohols, esters, anhydrides, lactones, nitriles, chlorinated hydrocarbons, tetra-

TABLE 4-2
Properties of unfilled perfluor-elastomer ECD-006 reinforced with carbon-black.[1,9]

Property	Physical unit	Values measured for the Elastomer	Elastomer + 10% carbon-black	Elastomer + 20% carbon-black
Density	g/cm^3	2.01	2.02	
Coefficient of linear thermal expansion	K^{-1}	3.2×10^{-4}	1.9×10^{-4}	
Glass temperature	°C	−12		
Brittleness temperature	°C			−39
Tensile strength	N/mm^2	10–18	15–20	19
Ultimate elongation	%	150–190	120–160	160
Modulus	N/mm^2	3.5–7.0	8.5–12.7	9.5
Compression set (70 h at 204°C)	%	25–35	28–50	34
Hardness (Shore A)	—	85	85	89
Dielectric constant	—	2.8–3.2		
Dissipation factor	—	0.001		
Volume resistivity	Ω cm	10^{18}		
Dielectric strength	kV/cm	> 180		
Refractive index	—	1.350		

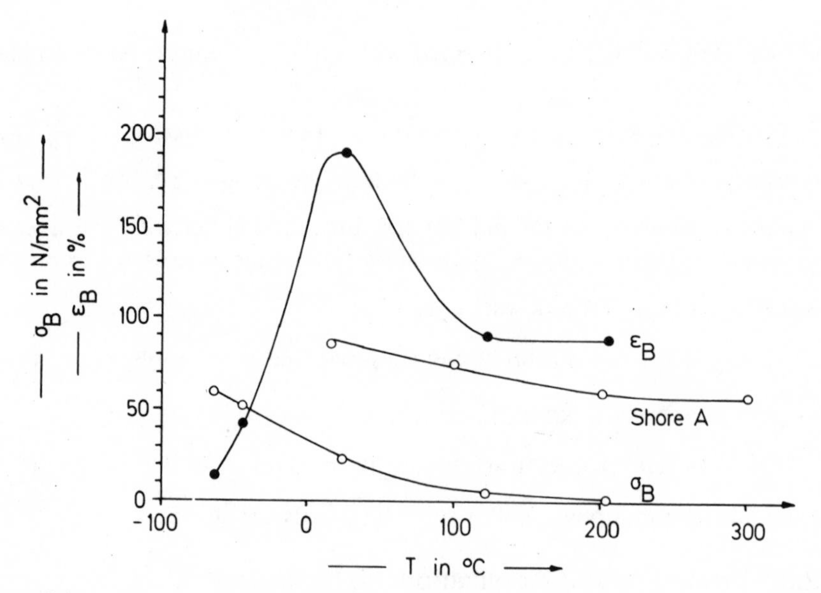

Figure 4-1 Temperature dependencies of tensile strength σ_B, ultimate elongation ε_B and Shore hardness (A) for a copolymer of tetrafluoroethylene and perfluoromethyl vinyl ether (ECD-006 from DuPont), reinforced with 20% carbon-black.[1,9]

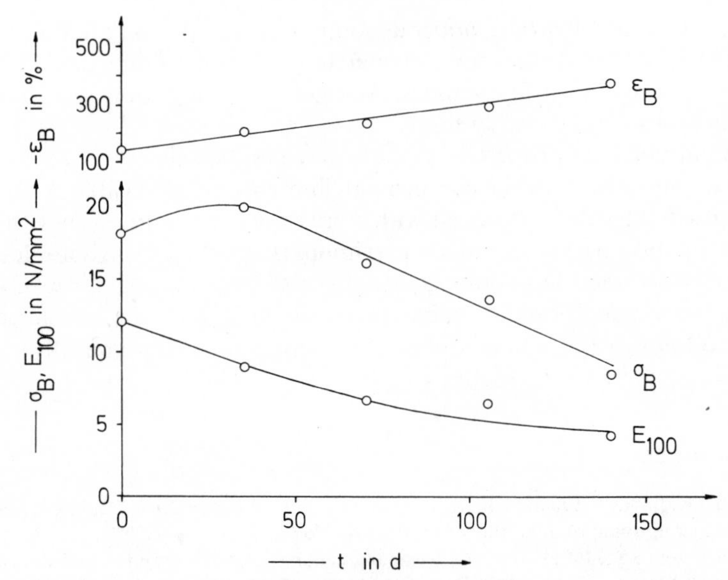

Figure 4-2 Time dependencies of tensile strength σ_B, ultimate elongation ε_B and 100%-modulus E_{100} for a copolymer of tetrafluoroethylene and perfluoro (methyl vinyl ether) (ECD-006 from DuPont), reinforced with 10% carbon-black. Measurements are for room temperature after aging in air at 288 °C for the times given.[2]

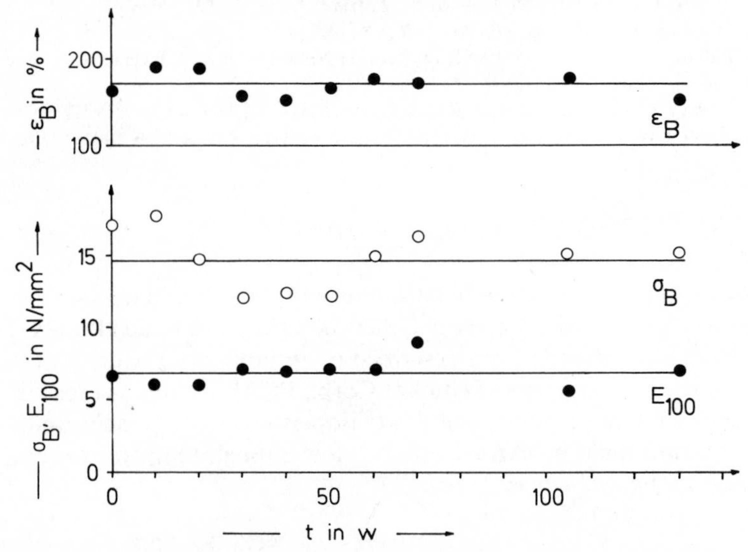

Figure 4-3 Time dependencies of tensile strength σ_B, ultimate elongation ε_B, and 100%-modulus E_{100} for a copolymer of tetrafluoroethylene and perfluoro(methyl vinyl ether) (ECD-006 from DuPont), reinforced with 10% carbon-black. Measurements are at room temperature after aging for the times given at 232°C in moist air under pressures of 0.04 bar.[1,2]

hydrofuran, nitrobenzene, aqueous sodium hydroxide and aluminum alkyls. On the other hand, the raw polymer is swollen by aldehydes, fluorinated hydrocarbons and a few amines. After cross-linking, only the fluorinated hydrocarbons still cause swelling.

The mechanical properties of these new perfluor-elastomers are comparable to those of existing commercial fluorelastomers (Table 4-2). Tensile strength and hardness decrease with increasing temperature, while the ultimate elongation passes through a maximum (Figure 4-1). Carbon-black-filled perfluorelastomers have quite good aging resistance in air (Figure 4-2). Under outer-space conditions (40 mbar moist air at 232°C) the tensile strength, ultimate elongation, and modulus of elasticity remained practically unaltered for almost three years (Figure 4-3).

References

1. G. H. Kalb, R. W. Quarles, Jr., and R. S. Graff, *Appl. Polymer Symp.*, **22**, 127, (1973).
2. Product information literature of DuPont de Nemours.
3. U.S. Patent 3 546 186; DuPont de Nemours; inv.: E. K. Gladding and R. Sullivan; see Ref. 4.
4. French Patent 1 527 816; Brit. Pat. 1 145 445 (March 12, 1969); DuPont de Nemours; inv.: D. G. Anderson, E. K. Gladding and R. Sullivan, *C. A.*, **70**, 107090b (1969).
5. U.S. Patent 3 467 638; French Pat. 1 555 360 (January 24, 1969); DuPont de Nemours; inv.: D. B. Pattison; *C. A.*, **71**, 40035n (1969).
6. G. H. Kalb, A. L. Barney and A. A. Khan, *Polymer Preprints*, **13**, 490 (1972).
7. D. C. Thompson, *Innovation* [*DuPont*], **4**(2), 8 (1973).
8. U.S. Patent 3 132 123 (May 5, 1964); DuPont de Nemours; inv.: J. F. Harris, Jr., and D. I. McCane; *C. A.*, **61**, 1968h (1964).
9. A. L. Barney, G. H. Kalb and A. A. Khan, *Rubber Chem. Technol.*, **44**, 660 (1971).
10. A. L. Barney, W. J. Keller and N. M. van Gulick, *J. Polymer Sci.* [*A*-1], **8**, 1091 (1970).

4.3 CARBOXYNITROSO RUBBER

The radical copolymerization of tetrafluoroethylene (I) and trifluoronitrosomethane (II) gives alternating copolymers comprised of I and II units. These copolymers have already been described in a number of review articles.[1-5] The commercial copolymers (Thiokol Corp., PCR) contain as cross-linking agent about 1 mol % additional 4-nitrosoperfluorobutyric acid units. The polymerization must be carried out at a low enough temperature to avoid formation of the cyclic oxazetidine:

$$\underset{\text{I}}{CF_2{=}CF_2} + \underset{\text{II}}{CF_3NO} \begin{cases} \xrightarrow{-20 \text{ to } 0^\circ C} \left(CF_2CF_2{-}N(CF_3){-}O\right) \\ \xrightarrow{100^\circ C} \begin{array}{c} CF_3{-}N{-}O \\ |\quad\;\; | \\ CF_2{-}CF_2 \end{array} \end{cases} \tag{4-6}$$

The high molecular copolymers have a very low glass temperature of $-50^{\circ}C$. They are stable to strong acids and oxidizing agents, even to pure oxygen itself. Carboxynitroso rubbers, in contrast to the other fluoroelastomers, are thermally unstable and are sensitive to bases.

The carboxynitroso polymers (CNR) may be vulcanized with metal oxides, diepoxides or by ligand exchange with chromium trifluoroacetate. The last of these systems appears to be favored by industry.[6] The CNR have a lower continuous service temperature of $-40^{\circ}C$, a tensile strength of 11 N/mm^2, an ultimate elongation of 165%, a compression set of 15–30% and a Shore hardness (A) of 74.[6] The electrical properties are: dielectric constant 2, volume resistivity 10^{15} Ω cm, dielectric strength 200 kV/cm and dissipation factor 0.005. A continuous service temperature of 190°C is given.[6]

These very cold-resistant elastomers are used, for instance, for jet aircraft under arctic and space conditions, for example as containers for rocket propellants (i.e. N_2O_4).

References

1. C. B. Griffis and M. C. Henry, *Rubber Chem. Technol.*, **39**, 481 (1966).
2. M. C. Henry, C. B. Griffis and E. C. Stump, *Fluorine Chem. Revs.*, **1**, 1 (1967).
3. J. Green, Nitroso Polymers, *Encycl. Polymer Sci. Technol.*, **9**, 322 (1968).
4. L. J. Fetters, in L. A. Wall (ed.), *Fluoropolymers* (High Polymers, Vol. 25), Wiley-Interscience, New York, 1972.
5. R. G. Arnold, A. L. Barney and D. C. Thompson, *Rubber Chem. Technol.*, **46**, 619 (1973).
6. Product information literature, PCR, Inc.

4.4 POLY(PERFLUOROETHYLENESULFONIC ACID)

4.4.1 Structure and Synthesis

The copolymerization of tetrafluoroethylene (I) and "sulfonylfluoride vinyl ether" (II) produces a copolymer which is named an XR-resin by DuPont de Nemours:

$$\underset{\text{I}}{CF_2{=}CF_2} \qquad \underset{\text{II}}{CF_2{=}CF{-}O{-}(CF_2{-}CF(CF_3){-}O)_n{-}CF_2CF_2SO_2F}$$

The commercial copolymer contains the two monomer units in a ratio of 5–10 mol I/1 mol II.[1,2] Saponification to the sulfonic acid of membranes, tubes etc. prepared from the XR-resin gives the fluoropolymer called Nafion®.

Monomer II is formed from the reaction of hexafluoropropylene oxide (III) with a sultone (IV). Hexafluoropropylene oxide is synthesized by the oxidation of hexafluoropropylene

$$\underset{\substack{|\\ CF_3}}{CF_2{=}CF} \longrightarrow \underset{\substack{|\\ CF_3}}{\overset{O}{CF_2{-}CF}} \quad \text{III} \tag{4-7}$$

and the sultone by

$$CF_2{=}CF_2 + SO_3 \longrightarrow \underset{\text{IV}}{\begin{array}{c} CF_2{-}CF_2 \\ |\quad\ | \\ O{-}SO_2 \end{array}} \xrightarrow{F^{\ominus}} \underset{\text{V}}{\overset{O}{\underset{F}{\diagdown}}C{-}CF_2{-}SO_2F} \tag{4-8}$$

The reaction of III and V leads to

$$z\,F_2C\overset{O}{\text{—}}\underset{\substack{|\\ CF_3}}{CF} + \overset{O}{\underset{F}{\diagdown}}C{-}CF_2SO_2F \rightarrow \overset{O}{\underset{F}{\diagdown}}C{\leftarrow}\underset{\substack{|\\ CF_3}}{CF}{-}O{-}CF_2{\rightarrow_z}CF_2{-}SO_2F \tag{4-9}$$

$$\overset{O}{\underset{F}{\diagdown}}C{-}\underset{\substack{|\\ CF_3}}{CF}{-}O{-}CF_2{\leftarrow}\underset{\substack{|\\ CF_3}}{CF}{-}O{-}CF_2{\rightarrow_{z-1}}CF_2SO_2F + Na_2CO_3 \longrightarrow$$

$$\longrightarrow CF_2{=}CF{-}O{-}CF_2{\leftarrow}\underset{\substack{|\\ CF_3}}{CF}{-}O{-}CF_2{\rightarrow_{z-1}}CF_2\,SO_2F + CO_2 + 2\,NaF \tag{4-10}$$

The copolymer which results from the copolymerization of II with I can be melted and can be processed by normal fabrication methods for thermoplastics. The copolymer is stable to water and acids, but is hydrolyzed to the corresponding sulfonic acid sodium salt with hot caustic. The sodium salt is then converted to the acid, with, for instance, nitric acid:

$$\sim CF_2SO_2F \xrightarrow{+\,NaOH} \sim CF_2SO_2Na \xrightarrow{+\,HNO_3} \sim CF_2SO_2H \tag{4-11}$$

4.4.2 Properties and Applications

Nafion® is offered as film in thicknesses between 0.05 and 0.25 mm as well as tubing or laminates. Water absorption depends, naturally, on the sulfonic acid group content, previous treatment and on the surrounding electrolytes. Nafion® is permselective; that is, it passes cations, but not anions. Exchange capacity is about 0.85 mmol/g. The tensile strength of resin containing 25% water is about 21 N/mm^2, the ultimate elongation is about 150 %. Strength is increased when the resin is reinforced with woven backing, for instance made of poly(tetrafluoroethylene).[3]

Nafion® is used as membranes for electrochemical processes, e.g. for the manufacture of sodium hypochlorite.

References

1. D. J. Vaughan, *Innovation* [*DuPont*], **4**(3), 10 (1973).
2. M. F. Hoover and G. B. Butler, *J. Polymer Sci.* [*Symp.*] **45**, 1, (1974).
3. W. Grot, *Chem.-Ing. Technik*, **44**, 167 (1972).
4. U.S. Patent 3282875 (November 1, 1966); Brit. Pat. 1034197 (June 29, 1966); DuPont de Nemours; inv.; D. J. Connolly and W. F. Gresham; *C. A.*, **66**, 11326z (1967).
5. U.S. Patent 3301893 (January 31, 1967); French Pat. 1406778 (July 23, 1965); DuPont de Nemours; inv.: R. E. Putnam and W. D. Nicoll; *C. A.*, **63**, 16216h (1965).
6. U.S. Patent 3560568; Dtsch. Off. 1959143 (July 23, 1970); DuPont de Nemours; inv.: P. R. Resnick; *C.A.*, **73**, 76659*z* (1970).
7. U.S. Patent 3692569 (September 19, 1972); DuPont de Nemours; inv.: W. G. F. Grot.
8. U.S. Patent 3718627 (February 27, 1973); DuPont de Nemours; inv: W. G. F. Grot.
9. U.S. Patent 3684747 (August 15, 1972); DuPont de Nemours; inv.: R. L. Coalson and W. G. F. Grot.

4.5 POLY(TRIFLUOROCHLOROETHYLENE-CO-ETHYLENE)

4.5.1 Structure

The E-CTFE-polymers produced by Allied Chemicals under the tradename Halar® are alternating copolymers of trifluorochloroethylene (I) and ethylene (II). The 1:1 mol ratio of the base units represents an 80 wt.-% content in trifluorochloroethylene.[1,2] 82% of the base units appear as alternating units, while 8% of the base units are found in ethylene blocks, and 10% are found in trifluorochloroethylene blocks.[3]

$$\underset{\text{I}}{CF_2{=}CFCl} \qquad \underset{\text{II}}{CH_2{=}CH_2}$$

4.5.2 Properties and Fabrication

The alternating units are present in an all-*trans* conformation. The mechanical properties are similar to those of polyamide 6. Halar® shows excellent impact strength (see Table 4-3). It has low creep, consequently, it is well-suited for sealants. The brittleness temperature is very low, so that the material can be used in the cold.

There are no solvents for it at temperatures up to +120°C. Halar is inert to acids, bases, strong oxidizing agents and most organic solvents. Polar organic

TABLE 4-3
Properties of Halar-polymers.[1-3]

Property	Physical unit	Values measured for Halar 300	Fiber (stretching ratio 5:1)
Density	g/cm^3	1.168	1.68
Melting temperature	°C	240–245	
Continuous service temperature	°C	150	
Heat distortion temperature (1.85 n/mm^2)	°3	78	
Brittleness temperature	°C	−80	
Tensile strength (23°C)	N/mm^2	49	
(−196°C)	N/mm^2	176	
Ultimate elongation	%	230	9
Tenacity	km		28
Tensile yield strength	N/mm^2	32	
Modulus (−196°C)	N/mm^2	6700	
Impact strength with notch (23°C)	—	no break	
Flexural modulus	N/mm^2	1700	
Creep modul (250 h at 10.6 N/mm^2)	N/mm^2	635	
Rockwell hardness	—	93	
Dielectric constant (60 Hz)	—	2.5	
Dissipation factor (60 Hz)	—	0.001	
Dielectric strength	kV/cm	800	
Arc resistance	s	135	
Limiting oxygen index (LOI)	%	60	48–50

solvents, when hot, swell Halar, but do not induce stress-corrosion. Water vapor permeability is as low as that of poly(vinylidene fluoride).

The powdered Halar can be used for coating processes, fluidized-bed and rotational molding. Granulated Halar can be used for injection molding, melt spinning and film splitting. These fibers are dyeable in green, blue and scarlet red shades. Fabrics from these fibers behave similarly to Nomex® relative to flammability and smoke generation. Use of these textiles as filter cloths and as upholstery for airplane seats is foreseen.

References

1. A. B. Robertson and W. A. Miller, *Chem. Engng.* (reprint without reference to volume, year and page).
2. A. B. Robertson and W. A. Miller, *Wire and Wire Products* (reprint without reference to volume, year and page).
3. A. B. Robertson, *Appl. Polymer Symp.*, **21**, 89 (1973).

4.6 POLY(VINYLIDENE FLUORIDE-CO-1-HYDROPENTAFLUOROPROPYLENE)

Copolymerization of vinylidene fluoride (I) with 1-hydropentafluoropropylene (II) results in copolymers which are marketed by the firm Montedison under the tradename Tecnoflon SL®.

$$\underset{\text{I}}{CH_2{=}CF_2} \qquad \underset{\text{II}}{CF_2{=}CH(CF_3)}$$

In addition to I and II, Tecnoflon T contains tetrafluoroethylene as well.[1,2] The composition of Types SH and FOR has not been divulged. However, type FOR is an unsaturated, cross-linkable fluoropolymer.[3]

Those properties of the Tecnoflon types which have been revealed are listed in Table 4-4. In general, types SL and T show somewhat poorer properties than do the copolymers of vinylidene fluoride and hexafluoropropylene (Viton® of DuPont, Fluorel® of 3M and SKF® of the USSR), especially with regard to solvent resistance.[4]

TABLE 4-4
Properties of Technoflon-resins.[3,4]

Property	Physical unit	Values measured for types		
		SL, SH	T	FOR
Density	g/cm^3			1.79
Heat distortion temperature	°C	260	280	
Continuous service temperature	°C	180	185	
Brittleness temperature	°C	−40	−41	
Tensile strength	N/mm^2	17.6	17.6–19	17
Ultimate elongation	%	160	200	200
Modulus (100%)	N/mm^2	7	4.6–5.1	5.8
Compression set (70 h at 200°C)	%	22	25–32	18
Shore hardness	—	75	70	69

References

1. U.S. Patent 3331823 (1967); Belg. Pat. 545894; Montecatini-Edison; inv.; D. Sianesi, G. C. Bernardi and A. Regio.
2. U.S. Patent 3335106 (1967); Dutch Pat. 6509095 (February 1, 1966); Montecatini-Edison; inv: D. Sianesi, G. C. Bernardi and G. Diotaller; *C. A.*, **65**, 903a (1966).
3. S. Geri, G. Giunchi and G. Ceccato, *Mater. Plast. Elastomeri*, **39**, 875 (1973).
4. R. G. Arnold, A. L. Barney and D. C. Thompson, *Rubber Chem. Technol*, **46**, 619 (1973).

4.7 POLY(PERFLUORO(ALKYL VINYL ETHER))

Since 1973, a so-called poly(perfluoroalkyl ether) has been marketed under the tradename Teflon PFA® (DuPont). It is a copolymer of tetrafluoroethylene (I) with a few percent of a perfluoro (alkyl vinyl ether) (II)

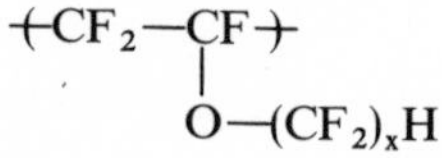

$$\left(CF_2 - \underset{\displaystyle O-(CF_2)_xH}{\underset{|}{CF}} \right)$$

The number x has not been revealed. Teflon PFA® is offered in two forms, each with differing melt indices (see Table 4-5).

TABLE 4-5
Properties of Teflon PFA.

Property	Physical unit	Values measured for type 9705 Refs. 1–3	9704 Refs. 3–5
Density	g/cm³	2.12–2.17	2.13–2.16
Coefficient of linear thermal expansion			
21–100°C	K^{-1}	1.2×10^{-4}	1.2×10^{-4}
100–149°C		1.7×10^{-4}	1.7×10^{-4}
150–209°C		2.0×10^{-4}	2.0×10^{-4}
Melting temperature	°C	305	302–310
Continuous service temperature	°C	260	260
Tensile yield strength 23°C	N/mm^2	15	14
250°C	N/mm^2	4.2	3.6
Tensile strength 23°C	N/mm^2	31	28
250°C	N/mm^2	14	13
Ultimate elongation 23°C	%	300	300
250°C	%	500	480
Flexural modulus 23°C	N/mm^2	703	670
250°C	N/mm^2	70	56
Folding endurance (0.02 cm)	—	500000	50000
Hardness (Shore D)	—	60	60
Dielectric constant ($60–10^8$ Hz)	—	2.1	2.06
2.4×10^{10} Hz	—		2.05
Dissipation factor $10^2–10^5$ Hz	—	0.002–0.003	0.00003
10^8 Hz	—		0.00045
10^{10} Hz	—		0.00131
Volume resistivity	Ω cm	$>10^{18}$	10^{18}
Dielectric strength	kV/cm	790	790
Flammability	—	flame retardant	
Limiting oxygen index (LOI)	%		>95
Water absorption	%	0.003	0.003

Teflon PFA® is a transparent plastic, which is fabricated as a thermoplastic. At the same time, however, it has the good thermal properties of poly(tetrafluoroethylene) which cannot be processed in the melt. As with other thermoplastics, fabrication can be carried out by injection molding or extrusion. The equipment must be corrosion proof, however, and must be able to withstand temperatures of 315–425°C. It must also be well ventilated.

Teflon PFA® has good stability to chemicals, a low coefficient of friction and very good electrical properties (Table 4-5). Thermal stability is relatively high after 3½ months at 285°C; there was still no drop found in the tensile strength at maximum load, the elongation and the flow properties.[2] in fact, tensile strength at maximum load and extensibility even increased with time (Figure 4-4).

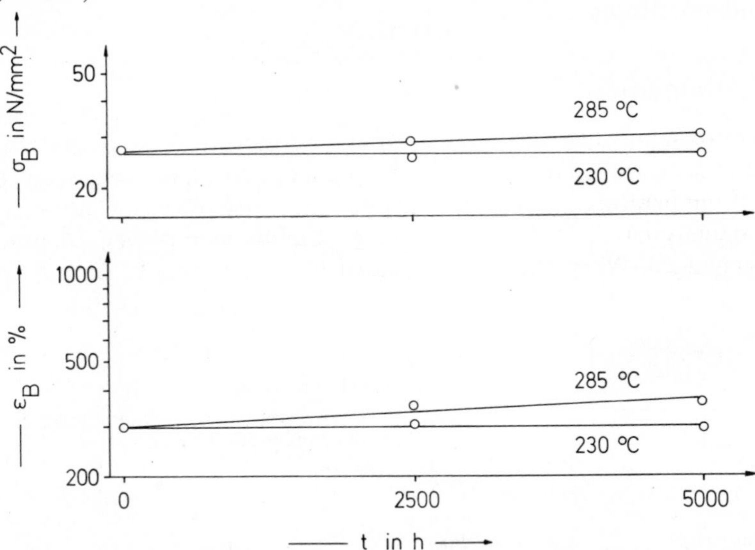

Figure 4-4 Dependence of tensile strength σ_B and ultimate elongation ε_B of Teflon PFA on aging time at 230 and 280°C. Measurements at room temperature.[5]

Teflon PFA® is principally recommended for use requiring prolonged exposure to elevated temperatures in the presence of chemicals, for instance, linings for valves, pipes, pumps and fittings, and for wire and cable sheathes.

References

1. Anonym., *Kunststoffe*, **64**, 92 (1974).
2. Product information literature of DuPont.
3. K. I. McCane, *Plastics Technology*, **19**(9), 18 (1973).
4. D. B. Allen, J. R. Perkins, Jr., J. C. Reed and E. W. Fasig, *Plastics Engng.*, **30**(10), 48 (1974).
5. D. B. Allen, J. R. Perkins, J. C. Reed and E. W. Fasig, *32nd Annual Techn. Conf., Soc. Plastics Engineers*, San Francisco, 1974.

4.8 POLY(HEXAFLUOROPROPYLENE OXIDE)

4.8.1 Synthesis and Structure

The polymerization of hexafluoropropylene oxide leads to "perfluoroalkyl polyethers" with the structure.

$$C_2H_5 \leftarrow\!\!\!\!\!\!+ OCF_2-CF(CF_3) +\!\!\!\!\!\!\rightarrow_n F$$

They are produced by DuPont and marketed under the tradename Krytox 143®. Krytox 240 is the name for a series of greases which consist of mixtures of Krytox 143-oils and Vydax® fluorotelomers. Vydax telomers are telomers of tetrafluoroethylene.

4.8.2 Properties

The Krytox family consists of colorless and odorless oils and greases. When heated under nitrogen or in air, the oils begin to decompose to gaseous products at temperatures over 360°C (pressure increase) or 470°C (differential thermal analysis). The properties of the remaining oil, however, are practically unchanged. When the oils are heated in the presence of certain metal

TABLE 4-6
Properties of Krytox oils and greases.

Property	Physical unit	Values measured for Krytox type				
		143 AZ	143 AB	143 AD	240 AZ	240 AD
Molecular weight	g/mol	2000		7000		
Viscosity, kin.						
−40°C	cm^2/s	80				
−18°C	cm^2/s	5.5	69			
38°C	cm^2/s	0.18	0.85	5.0	0.18	5.0
204°C	cm^2/s	0.008	0.018	0.06		
Pour point	°C	−57	−43	−29		
Boiling point (106 Pa)	°C	143–185	227–251	dec.		
Density 25°C	g/cm^3	1.86	1.89	1.91	1.89	1.93
204°C	g/cm^3	1.52	1.57	1.60		
Surface tension	N/cm	1.6×10^{-4}	1.9×10^{-4}	1.9×10^{-4}		
Dielectric constant	—	2.0	2.44			
Dielectric strength	kV/cm	153	203		>159	>159
Volume resistivity	Ω cm	6.4×10^{-15}	7×10^{-15}			
Dissipation factor	—	<0.003	<0.007	<0.007		

oxides, the decomposition temperature falls to 290° C. Heating the greases to temperatures over 200° C releases a small amount of toxic gases.

Krytox 143-oils and 240-greases are inert to boiling sulfuric acid, elemental fluorine, molten sodium hydroxide and, at room temperature, to ethanol, hydrazine, aniline, 90 % hydrogen peroxide and dinitrogen tetroxide. They are somewhat soluble in hydrazine and soluble to 25–35 % in dinitrogen tetroxide. The Krytox polymers do not react with liquid oxygen or with liquid dinitrogen tetroxide. However, explosion may result from contact with the surface of aluminum or magnesium cuttings. As long as the temperature remains under 95° C, the oils swell none of the elastomers, with the exception of natural rubber, *cis*-1,4-poly(butadiene) and styrene/butadiene copolymers. Only the fluorinated triazine elastomers withstand the oils at temperatures of 260° C. The greases are used as lubricants; the oils are used as lubricants, heat-transfer fluids, transformer fluids and as nonflammable oils for diffusion pumps. Typical properties are listed in Table 4-6.[1]

Reference

1. DuPont company publications Krytox G-5 and L-5.

4.9 POLY(HEXAFLUOROISOBUTYLENE-CO-VINYLIDENE FLUORIDE)

4.9.1 Structure and Synthesis

The Specialty Chemicals Division of Allied Chemical Corp. produces a CM-1® fluoropolymer, which is a 50/50 (mol/mol) copolymer of hexafluoroisobutylene (I) and vinylidene fluoride (II),[1] probably alternating:

$$CH_2{=}C(CF_3)_2 \qquad\qquad CH_2{=}CF_2$$

$$\text{I} \qquad\qquad\qquad\qquad \text{II}$$

The polymer is prepared by free radical suspension copolymerization under pressure by first adding liquid I to a water-containing reactor at 5° C followed by continuous addition of II at 20° C.[1,2] The polymer is available in limited quantities for selected application trials and development.[2]

4.9.2 Properties, Processing and Application

CM-1 has the same melting point as poly(tetrafluoroethylene) (327° C), but a lower density (1.88 vs. 2.17 g/cm^3). Catastrophic weight loss under nitrogen starts at 415°C. The continuous service temperature can be as high as 280°C, the heat distortion temperature (at 1.89 N/mm^2) 220°C. The coefficients of linear expansion are lower than those of poly(tetrafluoroethylene) and, furthermore, less temperature dependent: 3.7×10^{-5} K^{-1} (−45 to +24°C) and 4.7×10^{-5} K^{-1} (149 to 204°C). The linear mold shrinkage is only 0.02. CM-1 exhibits at 23°C a tensile strength of 39 N/mm^2, a tensile modulus of 3900 N/mm^2, and an elongation of 2%. Flexural strength and flexural modulus show similar values (see also Table 4-7).

TABLE 4-7
Properties of CM-1® Fluoropolymer.[2]

Property	Physical unit	Property value
Density	g/cm^2	1.88
Melting temperature	°C	327
(1.89 N/mm^2)	°C	220
Continuous service temperature	°C	280
Coefficient of thermal expansion		
(−45 to 24°C)	K^{-1}	3.7×10^{-5}
(149 to 204°C)	K^{-1}	4.7×10^{-5}
Tensile strength (23°C)	N/mm^2	37.9
(300°C)	N/mm^2	15.9
Tensile modulus (23°C)	N/mm^2	3790
(200°C)	N/mm^2	760
Elongation (23°C)	%	2
(300°C)	%	220
Flexural strength (23°C)	N/mm^2	36.5
Flexural modulus (23°C)	N/mm^2	4660
Impact strength	N	21.4
Hardness (Rockwell)	—	R 116
Limiting oxygen index	%	60
Relative permittivity (60 Hz)	—	2.30
(1 MHz)	—	2.25
Dissipation factor (60 Hz)	—	0.0015
(1 MHz)	—	0.0038
Volume resistivity (at 50% relative humidity and 23°C)	ohm cm	10^{17}
Dielectric strength (short time)	kV/cm	268
Water absorption	%	0.01
Critical surface tension	N/cm	1.93×10^{-4}

The polymer is inert to strong bases, halogens and metal salt solutions. The weight gain for CM-1 in ketones and esters was somewhat higher than for PTFE, but lower than that for poly(vinylidene fluoride) which was completely dissolved. None of the organics tested noticeably attacked or dissolved CM-1. CM-1 has also one of the lowest known water vapor permeabilities (four times lower than PTFE).

CM-1 fluoropolymer can be processed by compression molding (350°C), injection molding (370–380°C) or powder coating (fluidized bed, electrostatic spray, plasma spray). The polymers show good surface hardness (Rockwell R116), abrasion resistance, and scratch resistance. CM-1 adheres well to metals and has excellent nonsticking, nonwetting properties. CM-1 should thus have potential applications as mold release agent, nonstick coating for industrial rolls, cookware, chemical process equipment, fuel cells, or industrial fabrics. It is, thus, a strong competitor to poly(tetrafluoroethylene) for all applications except those which require a high impact strength.

References

1. F. Petruccelli, *ACS Coatings and Plastics Preprints*, **35**(2), 107 (1975).
2. Private communication, J. L. O'Toole, Allied Chem. Corp., October 29, 1975.

5 Polyethers

5.1 POLY(2, 6-DIPHENYL-1, 4-PHENYLENE OXIDE)

2,6-Diphenylphenol is available from cyclohexanone

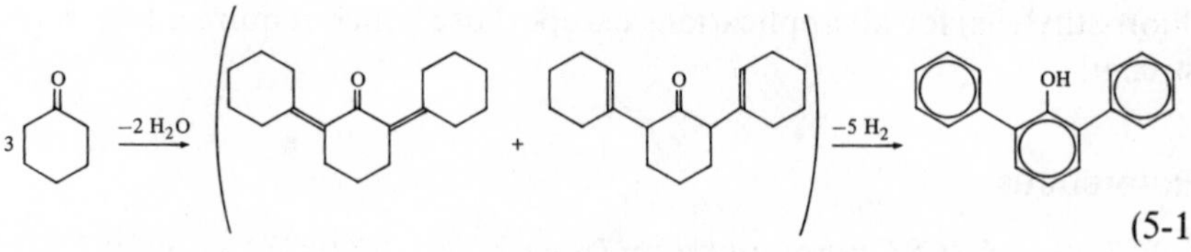

(5-1)

and is transformed into the corresponding polyether *via* oxidative coupling:

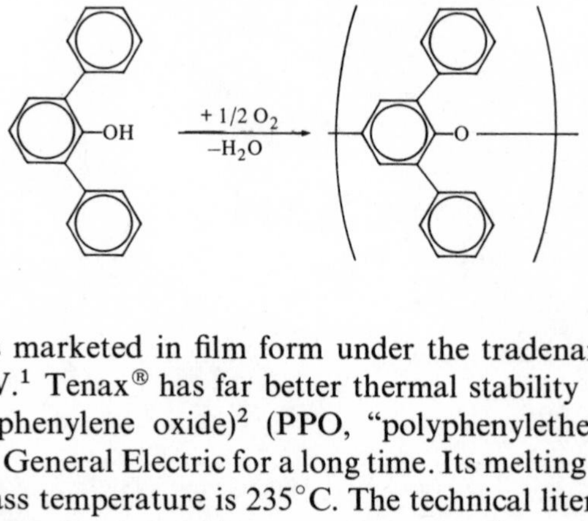

(5-2)

The polymer is marketed in film form under the tradename Tenax® by the firm Enka NV.[1] Tenax® has far better thermal stability than the poly-(2,6-dimethyl-1,4-phenylene oxide)[2] (PPO, "polyphenylether") which has been marketed by General Electric for a long time. Its melting temperature is 480°C, and its glass temperature is 235°C. The technical literature[1-3] does not disclose whether these transition temperatures refer to the α- or to the β-modification (see Ref. 4). Tenax® is stable in air up to 175°C.[1] Typical properties are listed in Table 5-1.

The high melting temperature of Tenax® precludes its use as a molding compound. The polymer is thus dry-spun from organic solvents. The filaments have a high degree of crystallization if stretched at high temperatures. Short fibers are fabricated into papers, which are used for extremely high voltage cable insulation.

TABLE 5-1
Properties of Tenax H-5 Films.[3]

Property	Physical unit	Values measured	
Thickness	μm	95±0.05	175±0.008
Density	g/cm³	0.75	0.83
Tensile strength			
(drawing direction)	N/mm²	25	29
(perpendicular to drawing dir.)	N/mm²	22	23
Modulus of elasticity			
(drawing direction)	N/mm²	1400	1600
(perpendicular to drawing dir.)	N/mm²	1300	1300
Compressive modulus	N/mm²	3	4
Friction coefficient		<0.60	<0.60
Dielectric constant		2.15	2.30
Dielectric dissipation factor		<0.0005	<0.0005
Dielectric strength	kV/cm	>1300	>1000
Water absorption			
(20°C, 65% relative humidity)	%	0.2	0.2

References

1. Anonym., *Mater. Plast. Elastomeri*, **10**, 788 (1973).
2. A. S. Hay, Aromatic polyethers, *Adv. Polymer Sci.*, **4**, 496 (1967); *Polymer Engng. Sci.*, **16**, 1 (1976).
3. Product information literature of the firm Enka NV.
4. J. Boon and E. P. Magré, *Makromol. Chem.*, **136**, 267 (1970).

5.2 HYDANTOIN-CONTAINING EPOXY RESINS

5.2.1 Structure and Synthesis

Early epoxy resins contained aromatic groups, and later types also had cycloaliphatic residues. A new class of epoxy resins containing hydantoin-groups has been developed by Ciba-Geigy AG since 1966.[1] These came on the market in 1973/1974. The structure, synthesis and application of these new epoxy-resins are covered in over 49 patents.

The Ciba-Geigy XB 2818®[2] and XB 2793®[3-5] types are pure hydantoin-resins, while the 2826®[3-5] and XB 2715a®[6] types contain conventional epoxy-resins in addition to the hydantoin-resins. Ciba-Geigy also markets a series of specially formulated products:

XB 2793

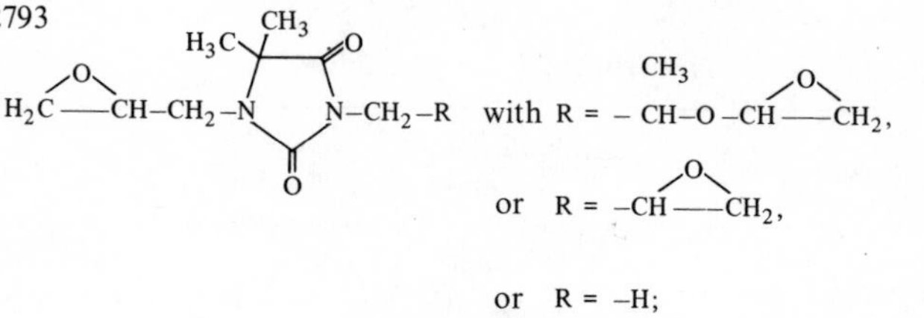

with R = $-CH(CH_3)-O-CH\overset{O}{—}CH_2$, I

or R = $-CH\overset{O}{—}CH_2$, Ia

or R = $-H$; Ib

XB 2818

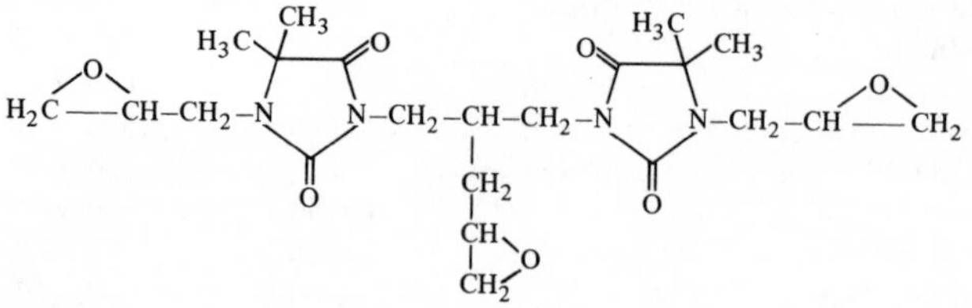

II

XB 2715a

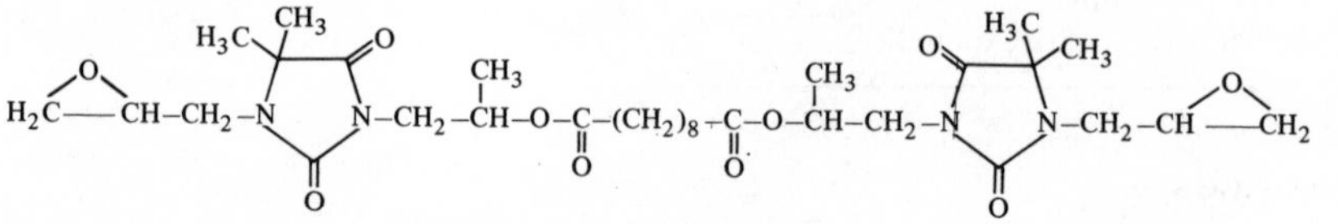

III

The synthesis of these resins can be carried out stepwise, since the hydantoin rings' two NH-groups have different reactivities: the one NH-group reacting more like an amide, the other reacting more like an imide

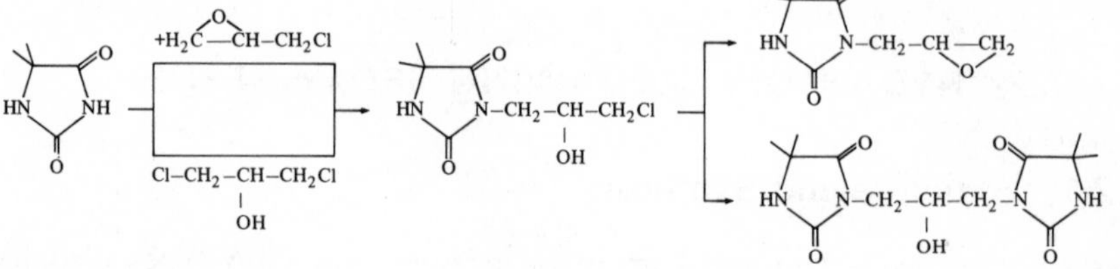

$+H_2C\overset{O}{—}CH-CH_2Cl$

$Cl-CH_2-CH(OH)-CH_2Cl$

(5-3)

The hydantoin-containing epoxy-resins are cured with cycloaliphatic dicarboxylic acid anhydrides

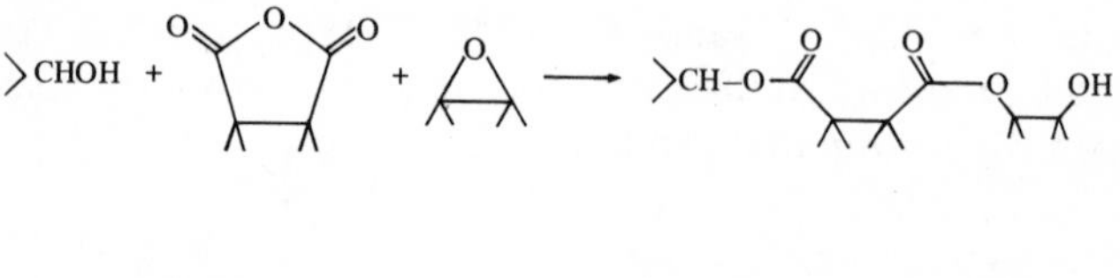

$>CHOH$ + + → $>CH-O-$ ... OH (5-4)

5.2.2 Properties and Application

The hydantoin resins are mobile to viscous fluids. They are offered principally in casting resin applications for the electrical and electronics industries. In comparison to conventional epoxy resins, they show higher heat distortion temperatures; they also accept more filler. In addition to the type of resin, the electrical and mechanical properties depend also significantly on the curing conditions. Typical approximate values are listed in Table 5-2.

TABLE 5-2
Properties of cured hydantoin-containing epoxy resins. Values are only approximate, since they depend strongly on cross-linking conditions.

Property	Physical unit	Values of I Ref. 1	Ia Ref. 7	II Ref. 2
Epoxide groups/mass	mol/kg	7.5		6.24
Anhydride/epoxide (mass/mass)		0.91		
Heat distortion temp.	°C			170
Softening temperature (Martens)	°C			159
Glass temperature	°C		130	
Flexural strength	N/mm^2	96	170	110–147
Impact flexural strength	kJ/m^2	51		118–137
Deflection	mm		11	4–6
Dielectric constant				3.6
Dissipation factor				0.008
Water absorption	%	0.54		0.5–0.6

References

1. D. Porret, *Makromol. Chem.*, **108**, 73 (1967).
2. J. Habermeier, *Angew. Makromol. Chem.*, **35**, 9 (1974).
3. U.S. Patent 3449353 (June 23, 1966); Ciba AG; Brit. Pat. 1148570 (April 16, 1969); Ciba Ltd.; *C. A.*, **71**, 3382c (1969).
4. Belgium Patent 759863 (December 5, 1969); Ciba AG.
5. Belgium Patent 750116 (January 20, 1970); Ciba AG.
6. British Pat. 1351525 (May 22, 1970); Ciba-Geigy AG; German Patent 2125355 (December 2, 1971); Ciba-Geigy AG; inv.: J. Habermeier, D. Baumann, D. Porret and H. Batzer; *C. A.*, **76**, 100615s (1972).
7. F. Lohse and R. Schmid, *Chimia [Aarau]*, **28**, 576 (1974).

5.3 PHENOLIC RESIN-FIBERS

5.3.1 Structure and Synthesis

Phenolic resins are the condensation products of phenols and aldehydes. In the base catalyzed curing of the initial products, the multiple functionality of the phenols leads to a cross-linked product, which, depending on reaction conditions, contains both dimethylene ether and methylene bridges, that is, schematically

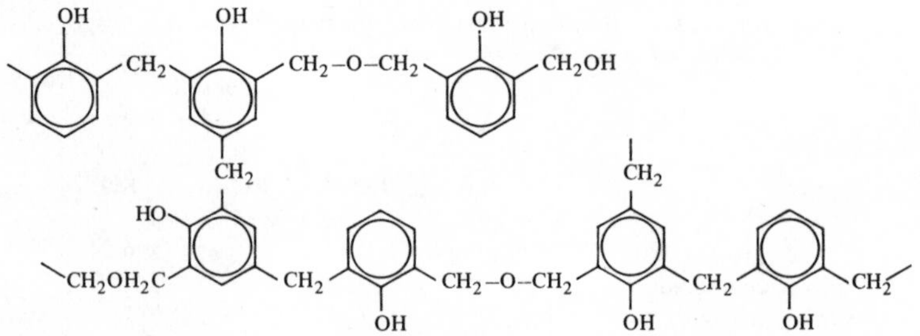

(5-5)

Phenolic resins are among the oldest synthetic polymers; however, the first fibers from phenolic resins were not reported until 1969! The synthesis of phenolic resin-fibers begins with a Novolak. Novolaks are produced from the acid-catalyzed condensation of formaldehyde with an excess of phenols. They have molecular weights of about 1000 g/mol and are soluble, that is they are not cross-linked. Spinning of one of these Novolaks in the melt at 130°C at about 200 m/min gives filaments which then are cured (cross-linked) with either gaseous formaldehyde at 100–150°C for 6–8 h or with a formaldehyde solution.

Phenolic resin-fibers are produced under the tradename Kynol® by American Kynol, Inc., a subsidiary of Carborundum AG. For review literature, see Ref. 2.

5.3.2 Properties and Application

The phenolic resin-fibers are yellow. However, acetylation of the phenolic hydroxyl groups gives white fibers.[3] The fibers have an elliptical cross-section with an axial ratio of 0.9. Phenolic resin-fibers are available in varying titers (see Table 5-3).

Phenolic resin-fibers are stable to many acids and bases at room temperature. An exception is nitric acid. Moisture absorption increases with increasing surface area, that is with decreasing titer. At room temperature and at

TABLE 5-3
Dependence of Kynol fiber properties on titer.

Diameter μm	Titer tex	Tenacity km	Elongation %	Modulus N/mm²
10–11	0.12	23	39	6300
11–15	0.12–0.22	21	35	5960
15–16	0.22–0.26	18	30	5630
16–17	0.26–0.29	17	25	5400
17–25	0.29–0.61	14	15	5000

approximately 65% relative humidity, for instance, it fluctuates between 4.2% for $\frac{4}{9}$ tex and 8.4% for $\frac{1}{9}$ tex.

The stress/strain behavior and the fracture work are strongly dependent on the fiber diameter[2] (Table 5-3). Thinner fibers are stronger and more extensible. Mechanical behavior also depends on the filament length, that is on filament defects: the shorter the fiber, the higher the tensile strength and the elongation. Further, the tensile strength of phenolic resin-fibers is greatest when the fibers are stretched at about 20–100%/min. The molecules are oriented during the stretching without crystallization. Consequently, the phenolic resin-fibers are amorphous. The amorphous structure appears to be responsible for the fact that, in spite of higher fracture work, the abrasion resistance is lower than that of Rayon. Typical property values are listed in Table 5-4.

The good stability of the phenolic resin-fibers is particularly noteworthy. Practically no weight loss is observed even after 1200 h at temperatures under

TABLE 5-4
Properties of Kynol Fibers.[2,4,5]

		Value measured			
		Unstretched			Stretched
Property	Physical unit	Untreated	500 h at 150°C	500 h at 260°C	
Density	g/cm³	1.25	—	—	—
Tensile strength	N/mm²	210	—	—	—
Tenacity	km	18	9	6	55
Ultimate elongation	%	20–35	5	3	<100
Modulus	N/mm²	6000	4600	5100	9000
Loop tenacity	N/mm²	110–300	—	—	—
Knot strength	N/mm²	150	—	—	—
Thermal conductivity	J/(h m K)	660	—	—	—
Dielectric strength	kV/cm	157	—	—	—

150°C in air. However, in the same time period the tensile strength did drop to $\frac{1}{3}$ of the starting value, the modulus fell approximately 25 %, and the elongation dropped abruptly from 34 to 2 %.[2] Oxidation occurs readily above 150° C Thermal stability can be raised from 150 to 200° C by acetylating the fibers. This also considerably improves light stability.[3]

The limiting oxygen index (LOI) is 36%. The fibers, therefore, do not burn, rather they only char with retention of shape in the flame. Shrinkage is less than 15%. The fact that these fibers do not melt on exposure to heat is important for their use as flame resistant fibers. In addition, there is little generation of smoke or toxic fumes, essentially only CO_2.

Phenolic fibers are offered as stable fibers, nonwoven fabrics, felt, and papers. They are used, for instance, for flame resistant blankets or for comfortable flame resistant occupational clothing. The fibers can be dyed with cationic or dispersion dyes, by which black, blue, green, red, orange and gold colors are obtainable. The relatively poor abrasion resistance can be improved by using mixed fiber blends with, for instance, wool or Nomex®.

References

1. J. Economy, L. Wohrer, F. Frechette, Kynol—A new flame resistant fiber, 39th Annual Meeting Textile Research Institute, New York, April 1, 1969.
2. J. Economy and L. Wohrer, Encycl. *Polymer Sci. Technol.*, **15**, 365 (1971).
3. J. Economy and L. C. Wohrer, F. J. Frechette and G. Y. Lei, *Appl. Polymer Symp.*, **21**, 81 (1973).
4. Product information literature of American Kynol, Inc.

5.4 ARALKYL PHENOLIC RESINS

Albright and Wilson Ltd., England, is manufacturing a series of thermosetting resins which are marketed by Ciba-Geigy under the tradename Xylok®. The polymers are condensation products of aralkyl ethers and phenols, e.g. synthesized from α,α′-dimethoxy-*p*-xylene and phenol under the action of Friedel-Crafts catalysts

OH OH OH

$$-CH_2-\quad -CH_2-CH_2- \quad]_n \quad -CH_2- \quad -CH_2-$$

Xylok 210 is used for the production of composites with glass, asbestos, and carbon fibers, Xylok 225 for the manufacture of filled compression, transfer, and injection grade molding compounds. All resins are easy to process. Xylok 210 constitutes a one-pack resin system: no separate catalyst is needed for curing. Xylok 225 is cured by hexamine. Postcuring is essential for both types of resins.

Some properties of Xylok resins are collected in Table 5-5. Xylok 210 shows good thermal stability up to 250°C, good mechanical strengths at room and elevated temperatures, and excellent dielectric properties. Xylok 225 also exhibits good thermal stability and good mechanical properties but, in addition, low water absorption and good wear resistance.

Xylok 210 composites are used for bearings, friction linings, grinding wheels, lamp capping cements, insulation of coils, windings, copper conduc-

TABLE 5-5
Properties of aralkyl phenolic resins.

		Value for			
		Xylok 210		Xylok 225	
Property	Physical unit	Glass-cloth[a] laminate	Asbestos laminate	Asbestos flour	Carbon fiber
Density	g/cm^{-3}	1.77	1.65	1.68	1.34
Shrinkage	%			0.004	0.004
Deformation temperature ($1.86 N/mm^2$)	°C	>330	325		
Tensile strength (23°C)	N/mm^2	443	60	53	42
(250°C)	N/mm^2	307	54[b]		
Tensile modulus	N/mm^2	38,000	11,000		
Elongation at break	%		0.62		
Crushing strength	N/mm^2	450	264		
Izod impact strength	N/cm	1010		15	21
Vickers hardness	—		36.6		
Interlaminar shear strength	N/mm^2	89			
Flexural strength (23°C)	N/mm^2			70	70
(250°C)	N/mm^2			58	49
Compressive strength	N/mm^2			253	295
Water absorption (24h, 100°C)	%	0.65			
Flammability		self-extinguishing			
Oxygen index (LOI)	%	55			
Coefficient of friction	—	0.215			
Electric strength (90°C)	kV/cm				102
Relative permittivity (1 MHz)	—			4.8	
Power factor	—			0.053	
Volume resistivity	—			2×10^{-14}	

[a]Post-cured at 250°C.
[b]200°C.

tors and wires. Xylok 225 found application as engineering components and bearing materials, for domestic appliances and ablative linings, in the electrical and electronic industries, and as binder.

Literature

Company literature of Ciba-Geigy Corporation.

5.5 POLY[(1, 2-DI(CHLOROMETHYL) ETHYLENE OXIDE]

A new stereoregular polyether with the monomeric unit (I)

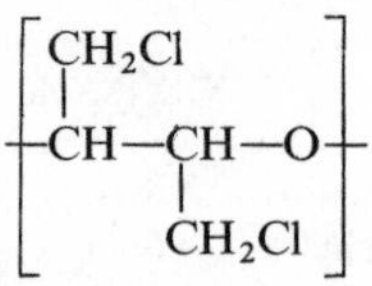

$$\left[\begin{array}{c} CH_2Cl \\ | \\ -CH-CH-O- \\ \quad\ | \\ \quad\ CH_2Cl \end{array}\right] \qquad \left[\begin{array}{c} CH_2Cl \\ | \\ -CH_2-C-CH_2-O- \\ | \\ CH_2Cl \end{array}\right]$$

I (*cis* or *trans*) 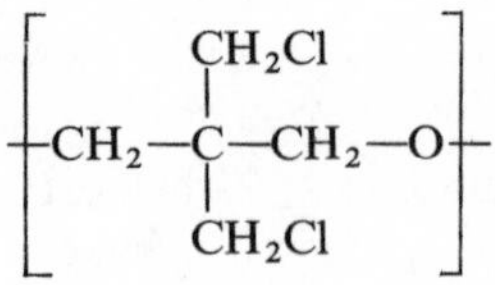 II

was recently announced by Hercules Inc.[1] It is possibly a replacement for Penton®(II) which was phased out by Hercules recently.

The monomer synthesis follows the following route: chlorine and butadiene react to give 1,2-dichlorobutene-3 (III) and 1,4-dichlorobutene-2 (IV) in roughly equal amounts:

$$2CH_2{=}CH{-}CH{=}CH_2 + 2Cl_2 \rightarrow \underset{III}{ClCH_2{-}CHCl{-}CH{=}CH_2} + \underset{IV\ (cis\ or\ trans)}{ClCH_2{-}CH{=}CH{-}CH_2Cl}$$

III is the precursor for chloroprene. IV is epoxidized *via* the acetic acid route to give 1,4-dichloro-2,3-epoxybutane (V) which is then polymerized by various catalysts such as $R_3Al/0.5\ H_2O$, R_2Zn/H_2O, and R_2Mg/H_2O to I:

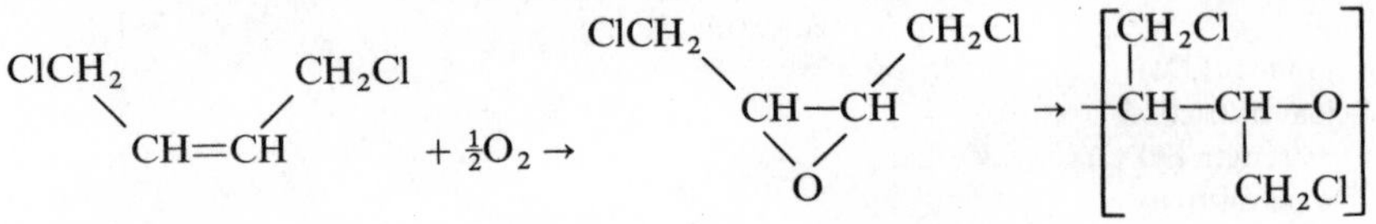

$$\underset{IV}{ClCH_2{-}CH{=}CH{-}CH_2Cl} + \tfrac{1}{2}O_2 \rightarrow \underset{V}{ClCH_2{-}\overset{O}{\widehat{CH{-}CH}}{-}CH_2Cl} \rightarrow \underset{I}{\left[\begin{array}{c} CH_2Cl \\ | \\ -CH-CH-O- \\ \quad\ | \\ \quad\ CH_2Cl \end{array}\right]}$$

The *cis*-polymer crystallizes in all-*trans* conformation (zig-zag) and exhibits a melting point of 235°C and a melting temperature of about 95°C. The *trans*-polymer forms a 4_1-helix with a melting temperature of 145°C; the glass temperature is about 55°C. The elongations are 2% (*cis*) and 3% (*trans*), the tensile strengths 36 (*cis*) and 50 N/mm^2 (*trans*), and the tensile modulus 2300 N/mm^2 (both). The impact strength is about half of that of poly(styrene), the thermal stability about 100°C above poly(vinyl chloride). The poor impact strength can be overcome by orientation. The electrical behavior resembles that of poly(ethylene terephthalate).

The polymer is probably going to be used as an engineering plastic, possibly also as a rubber-modified impact grade. The fiber properties are said to be similar to wool. The polymer also possesses good barrier properties for air and water.

References

1. E. J. Vandenberg, *J. Polymer Sci.* [*Chem.*], **13**, 2221 (1975).

6 Polyesters

6.1 POLY(PIVALOLACTONE)

6.1.1 Structure and Synthesis

Poly(pivalolactone) is the polyester of hydroxypivalic acid (β-hydroxy-α,α-dimethylpropionic acid, 2-hydroxy-1,1-dimethylpropionic acid)

$$-\!\!(O-CH_2-C(CH_3)_2-CO)\!\!-$$

Poly(pivalolactone) is produced in research quantities by the Royal Dutch Shell group.[1,2] Research is also apparently being carried out by Dynamit Nobel.[3] Hydroxypivalic acid does not polycondense to high molecular weights. "Steric hindrance" is given as the reason. The real reason is apparently impurities, either in the monomer or from decomposition reactions during the melt polymerization. Therefore, poly(pivalolactone) is produced by the polymerization of α,α-dimethyl-β-propiolactone (I), which in turn is synthesized from pivalic acid by chlorination followed by ring-closure of the chlorinated compound:[1]

$$(CH_3)_3CCOOH \xrightarrow[-HCl]{+Cl_2} (CH_3)_2C(CH_2Cl)COOH \xrightarrow[-H_2O]{+NaOH}$$

$$\xrightarrow[-H_2O]{+NaOH} ClCH_2-\overset{\overset{CH_3}{|}}{\underset{\underset{CH_3}{|}}{C}}-COONa \xrightarrow{-NaCl} CH_3-\overset{\overset{CH_3}{|}}{\underset{\underset{H_2C\,-\,O}{|\quad\quad|}}{C\,-\,C{=}O}} \qquad (6\text{-}1)$$

The polymerization is carried out with tributylphosphine as initiator with ring-opening to give a living zwitterion:

$$Bu_3P + CH_3-\overset{\overset{CH_3}{|}}{\underset{\underset{H_2C\,-\,O}{|\quad\quad|}}{C\,-\,C{=}O}} \longrightarrow Bu_3\overset{\oplus}{P}-CH_2-C(CH_3)_2-COO^{\ominus} \longrightarrow$$

$$\longrightarrow Bu_3\overset{\oplus}{P}-CH_2-C(CH_3)_2-(COO-CH_2-C(CH_3)_2)_n-COO^{\ominus} \qquad (6\text{-}2)$$

Apparently the growing zwitterions are present as intra- and/or intermolecular associates. Termination reactions can occur at elevated temperatures, that is, for instance, under extrusion conditions:

$$Bu_3\overset{\oplus}{P}\sim\sim COO^{\ominus} + Bu_3\overset{\oplus}{P}\sim\sim COO^{\ominus} \longrightarrow \begin{cases} Bu_3\overset{\oplus}{P}\sim\sim COOBu + Bu_2P\sim\sim COO^{\ominus} \\ Bu_3\overset{\oplus}{P}\sim\sim COBu + Bu_2\overset{\overset{O}{\|}}{P}\sim\sim COO^{\ominus} \end{cases} \qquad (6\text{-}3)$$

These termination reactions cap the living chain-end and, therefore, stabilize the polymer. The polymer begins to decompose by depolymerization at temperatures above the melting temperature (245°C) of the polymer. Depolymerization results in formation of pivalolactone, which in turn decomposes to isobutylene and carbon dioxide.

6.1.2 Properties, Fabrication and Application

Three crystalline modifications are observed with poly(pivalolactone). Principally the α-modification crystallized rapidly from the melt. In the α-modification the polymer chains are present in a helical conformation. The β-modification crystallizes significantly slower than the α-modification. On annealing at temperatures above its melting point of 228°C it changes over into the α-modification. The γ-modification arises on stretching the polymer. In it the polymer chains are present in the all-*trans* conformation, that is as zig-zag chains. On annealing the γ-modification is transformed into the α-modification accompanied by volume contraction of the polymer. As a consequence of the high purity of the polymer, only a few spherulites are formed on slow cooling of the melt. However, these spherulites are up to 1 mm in size. Spherulite formation can be repressed by the addition of nucleating agents, and in the case of very thin objects (fiber), also by very rapid cooling.

The density of the polymer depends on its degree of crystallinity. For material, determined to be 100 % crystalline by X-ray, it is 1.223 g/cm^3 at 20°C; for fully amorphous products it is 1.097 g/cm^3. Normal cooling in the mold leads to a density of 1.19 g/cm^3, or a degree of crystallinity of about 75 %.

Poly(pivalolactone) can be made into synthetic fibers, films and molded objects. Melt-spinning can be carried out at temperatures between 265 and 300°C. The polymer must be brought to temperature relatively fast, otherwise there is substantial decomposition to gaseous products. The rate of cooling must be carefully controlled because of the polymer's high rate of crystallization. Spherulite formation tendencies can be reduced through the addition of crystal nucleating agents, high rates of spinning, small filament diameters and the quenching of filaments in water. Still, some very fine spherulites form even under rapid cooling. However, these spherulites disappear on careful stretching on warmed plates or spindles at 50–210°C. If the fiber is first stretched, then annealed and finally stretched again, fibers are obtained with a high proportion of the γ-modification having a tenacity of 75–90 km (75–90 g/tex), a very high modulus (up to 1100 g/tex) and a very low ultimate elongation of 8–10%. The γ-modification disappears after annealing at 150–200°C. Annealed fibers contain the α-modification and show a tenacity reduced to 45–55 g/tex, a modulus lowered to 350 g/tex and an ultimate elongation increased to 60–80%. The fibers have a seemingly high elastic recovery of 80–90%, which, in fact, is largely independent of heat treatment in water vapor or in dry air.[2]

Poly(pivalolactone) is stable to acids, alkalis, solvents, bleaches and detergents. Weatherability is equally good, presumably because of its resistance to saponification. The fibers can be colored only with dispersion dyes, and then only with difficulty.

Poly(pivalolactone) must be treated with nucleating agents for the produc-

TABLE 6-1
Properties of Poly(pivalolactone).[2]

Property	Physical unit	Values measured: Thermoplastic, Original	Thermoplastic, Annealed[a]	Fiber, Original	Fiber, Stretched[b]
Density	g/cm^2	1.18			
Melting temperature	°C	245			
Glass temperature	°C	−10			
Vicat softening temperature	°C	220			
Heat distortion temperature (at 1.86 N/mm^2)	°C	180–200			
Tensile yield strength	N/mm^2	39	38		
Tensile yield at maximum elongation	%	4–6			
Ultimate elongation	%	10	69	400	28
Compression set at break	%	1.7	1.0		
Tenacity	km			10.8	27
Impact strength	kJ/m^2	4.6	40		
Tensile impact strength	KJ/m^2	100	7600		
Notched impact strength	kJ/m^2	3.4	6.5		
Torsion modulus, 1 Hz, 20°C	N/mm^2	1500			
100°C	N/mm^2	300			
Lengthening during creep experiment (10,000 h at 20°C and 20.6 N/mm^2)	%	0.8			
Shore Hardness (D)	—	85			
Water absorption (24 h)	%	0·2			
Volume resistivity	Ω cm	6×10^{13}			

[a]5 min at 200°C
[b]Stretching ratio 1:3.8 at 2.5 tex.

tion of molded articles. These molded parts then show good thermal and dimensional stabilities. Annealing significantly increases the ultimate elongation, while the tensile yield strength remains practically constant. Annealing apparently removes stresses introduced by crystallization during molding. This is supported by the fact that the various impact strength values were improved by annealing (Table 6-1).

References

1. N. R. Mayne, *Chem. Tech.*, **2**, 728 (1972); *ibid.*, in N. A. J. Platzer (Ed.); Polymerization reactions and new polymers, *Adv. Chem. Ser.*, **129** (1973).
2. H. A. Oosterhof, *Polymer* [*London*], **15**, 49 (1974).
3. N. Volkommer and G. Bier, *Angew. Chem.*, **86**, 448 (1974); *Angew. Chem. Internat. Ed. Engl.*, **13**, 412 (1974) (summary of a lecture).

6.2 HOMO- AND COPOLYMERS OF POLY-(*p*-HYDROXYBENZOATE)

6.2.1 Structure and Synthesis

Poly(*p*-hydroxybenzoates) contain the base unit I. A series of commercially available products are copolymers of I-base units with isophthalic acid base units (II), hydroquinone base units (III), terephthalic acid base units (IV) and *p,p'*-diphenyl ether base units (V):

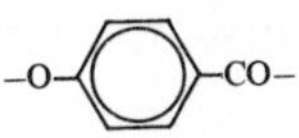

–O–⟨ring⟩–CO– I

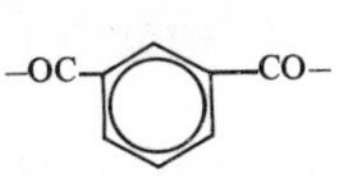

–OC–⟨ring⟩–CO– II

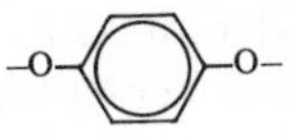

–O–⟨ring⟩–O– III

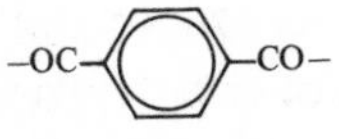

–OC–⟨ring⟩–CO– IV

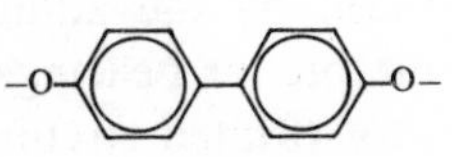

–O–⟨ring⟩–⟨ring⟩–O– V

All of these polymers are marketed by the firm Carborundum under various tradenames: Ekonol P-3000® is the unipolymer of I.[1–5] Ekonol T-4000® contains in addition to P-3000® also poly(tetrafluoroethylene). Other types are reinforced with glass fibers, molybdenum disulfide or graphite. In addition, aqueous dispersions of mixtures of Ekonol and poly(tetrafluoroethylene) are available (D-5000®). Ekkcel C-1000® is a copolymer of I, II and III,[6,7] while Ekkcel I-2000 is a copolymer of I, IV and V.[6–8]

The direct polycondensation of *p*-hydroxybenzoic acid at the required temperatures of 200–220° C leads to decarboxylation; consequently, not the acid, but rather its phenyl ester is polycondensed commercially.[6] Polycondensation is carried out in a 40/60 mixture of *o*- and *m*-terphenyl, because the presence of a good heat transfer agent is absolutely necessary for high conversions. Very slow heating (about 12 h) to the polycondensation temperature of 340° C is also important:

$$HO-C_6H_4-COOC_6H_5 \longrightarrow \text{+}O-C_6H_4-CO\text{+} + C_6H_5OH \qquad (6\text{-}4)$$

6.2.2 Properties, Fabrication and Application

The homopolymer is a white, highly crystalline material with a molecular weight of about 8000–12,000 g/mol. The melting temperature listed in Table 6-2[6,9] represents a minimum value. A sharp endotherm appears at 360°C, which is not attributed to melting, but rather to a loss of one dimensional order.[6]

The homopolymer is insoluble in dilute acids and bases and in all solvents up to their boiling points. Chlorinated biphenyls swell it somewhat at 325° C. Hot concentrated sulfuric acid and sodium hydroxide attack it. The polymer is extraordinarily thermally stable. Weight loss after 2000 h at 260° C is only 1 % and after 1 h at 400° C also only 1 %. At 480° C the weight drops at about 5 %/h. The limiting oxygen index (LOI) of the copolymers is about 37 %.

The homopolymer has metal-like properties (see Table 6-2). Like metals it can be shaped by hammering. Other fabrication methods are impact molding, plasma spraying, and sintering under pressure in molds at temperatures of 420° C and under pressures of 35 N/mm^2. Of course, the Ekkcel series of copolymers are easier to fabricate.

The self-lubricating properties already present in the homopolymer can be varied by the addition of graphite, boron nitride, molybdenum disulfide or poly(tetrafluoroethylene). Abrasives can be produced by the addition of silicon carbide. The homo- and copolymers are suitable for bearings, parts for data-processing systems, gear wheels, supports for printed circuits, high temperature electrical and electronic components and for cooking utensils.

References

1. German Patent 1720440; French Pat. 1568152 (May 23, 1969); Carborundum Corp.; inv.: J. Economy and B. E. Nowak; *C. A.*, **72**, 13412x (1970).
2. Anonym., *SPE-J.*, **26**(4), 33 (1970).
3. Anonym., *Kunststoffe*, **61**, 36 (1971).
4. J. Economy, B. E. Nowak and S. G. Cottis, *ACS Polymer Preprints*, **11**(1), 332 (1970).
5. J. Economy, B. E. Nowak and S. G. Cottis, *SAMPE J.*, **6**(6), 21–7 (Aug/Sept. 1970).

TABLE 6-2
Properties of homo- and copolymeric poly(hydroxybenzoates).

Property	Physical unit	Values measured for Ekonol P-3000 Refs. 2, 3, 5 and 9	Ekkcel C-1000 Refs. 6 and 10	Ekkcel I-2000 Refs. 6 and 10
Density	g/cm^3	1.45	1.35	1.40
Melting temperatures	°C	> 550	370	413
Glass temperature	°C		159	
Heat distortion temperature (at 1.86 N/mm^2)	°C		300	293
Coefficient of linear thermal expansion	K^{-1}	1.5×10^{-5}	5.2×10^{-5}	2.9×10^{-5}
Tensile strength, 23°C	N/mm^2		70	99
260°C	N/mm^2		21	21
Ultimate elongation	%		7–9	8
Modulus	N/mm^2	7250	1330	2550
Flexural strength, 23°C	N/mm^2	75	106	120
260°C	N/mm^2		35	28
Flexural modulus, 23°C	N/mm^2	510	3200	4900
260°C	N/mm^2		880	1410
Compressive strength	N/mm^2		140	127
Compressive modulus	N/mm^2		2100	3500
Impact strength	N		109	163
Notched impact strength	N		22	55
Rockwell hardness			124	88
Dielectric constant (1 kHz)		3.28	3.68	3.16
Dielectric strength	kV/cm	264	177	
Dissipation factor (1 kHz)		0.0025	0.0085	
Water absorption (24 h)	%	0.02	0.040	0.025
(500 h at 100°C)	%	0.4		

6. R. S. Storm and S. G. Cottis, *ACS Coatings and Plastics Preprints*, **34**, 194 (1974); *ibid.*, in R. D. Deanin, ed., *New Industrial Polymers, ACS Symp. Ser.*, **4**, 156 (1974).
7. U.S. Patent 3637595 (January 25, 1972); Carborundum Corp.; inv.: S. G. Cottis, J. Economy and B. E. Nowak.
8. Advertisement in *Plastics Eng.*, **30**(6), 69 (June 1974).
9. J. Economy and S. G. Cottis, *Encycl. Polymer Sci. Technol.*, **15**, 292 (1971).
10. Product information literature of the Carborundum Corp.

6.3 POLY(ETHYLENEOXY BENZOATE)

Poly(ethyleneoxy benzoate) with the base unit

$$-O-CH_2-CH_2-O-C_6H_4-CO-$$

has been known since 1946.[1] The polymer first gained commercial importance, however, when the firm Unitika (Nippon Rayon) managed to produce it in high enough molecular weight. Fibers of the polymer were manufactured and marketed between 1968 and 1976 under the tradename A-Tell® jointly by Unitika/Mitsubishi Chemical Industry.[2]

Phenol, carbon dioxide and ethylene oxide are starting materials for its synthesis:[3]

$$HO-C_6H_5 + CO_2 \xrightarrow{KOH} HO-C_6H_4-COOH \xrightarrow{+C_2H_4O}$$

$$\xrightarrow{+C_2H_4O} HOCH_2CH_2O-C_6H_4-COOH \xrightarrow{+CH_3OH} HOCH_2CH_2O-C_6H_4-COOCH_3$$

$$HOCH_2CH_2O-C_6H_4-COOCH_3 \xrightarrow{-CH_3OH} +OCH_2CH_2O-C_6H_4-CO+ \qquad (6\text{-}4)$$

The white, crystalline polymer has two crystal modifications, the relative amounts of which depend on the degree of stretching. The density of the α-form is 1.386 g/cm^3, the density of the amorphous polymer is 1.32 g/cm^3. The commercial polymer is about 41% crystalline and has a density of 1.34 g/cm^3. The melting point is 223°C, the glass temperature is 65°C (completely amorphous) or 84°C (41% crystalline).

The polymer is spinnable from the melt at 250–270°C. Crystallization occurs slowly, and consequently, the fiber shrinks on heating in the stretched state. Therefore, after stretching, the fiber is heated under tension to 150–180°C in order to accelerate crystallization and to improve thermal stability.

Fiber properties (see Table 6-3) lie somewhere between those of the harder poly(ethylene terephthalate) and the softer polyamide 6, that is, the fiber has silk-like properties. Textiles have good wrinkle-resistance, so that ironing is not necessary. The weatherability is somewhat better than that of polyamide 6 and poly(ethylene terephthalate). The fibers from poly(ethyleneoxy benzoate) can be dyed to deep shades with dispersion and azo dyes.

TABLE 6-3
Properties of poly(ethyleneoxy benzoates).

Property	Physical unit	Values measured for: Fibers, dry	Fibers, wet	25 μm films, biaxially stretched
Tensile strength	N/mm^2			180
Tenacity	km	36–48	36–38	
Ultimate elongation	%	15–30	15–30	127
Elastic recovery				
after 3% extension	%	95–100		
after 10% extension		75		
Modulus	N/mm^2	5000–9000		
Moisture content (20°C, 65% rel. humidity)	%	0.4–0.5		
Transparency (560 nm)	%			92
Gas permeability N_2	$cm^2\ s\ g^{-1}$			2.7×10^{-9}
O_2	$cm^2\ s\ g^{-1}$			95.6×10^{-9}
CO_2	$cm^2\ s\ g^{-1}$			203.4×10^{-9}

References

1. British Patent 604985 (July 14, 1948); Imperial Chem. Ind.; inv.: J. G. Cook, J. T. Dickson, A. R. Lowe and J. R. Whinfield; *C. A.*, **43**, 1223g (1949).
2. Anonym., *Chem. Engng. News*, 10 (Nov, 10, 1975).
3. K. Mihara, *Angew. Makrol. Chem.*, **40/41**, 41 (1974).

6.4 POLY(BUTYLENE TEREPHTHALATE)

Poly(butylene terephthalate) (PBT) or poly(tetramethylene terephthalate) (PTMT) is the polyester formed from 1,4-butanediol and terephthalic acid with the base unit

$$-O-(CH_2)_4-O-OC-C_6H_4-CO-$$

Poly(butylene terephthalate) has been brought on the market by at least ten companies since the expiration of the Whinfield–Dickson patents (see Table 6-4). The poly(butylene terephthalates) are engineering plastics which can be fabricated at significantly lower temperatures than poly(ethylene terephthalate). However, this advantage is obtained at the expense of a lowered glass temperature (see Table 6-5) and somewhat poorer mechanical

TABLE 6-4
Tradenames and types of poly(tetramethylene terephthalates).

Company	Tradename	Types		
		unreinforced	reinforced with 30% glass fiber	reinforced and with flame-retardant
BASF	Ultradur	B 4500		
Celanese	X 917			
Ciba-Geigy	Crastin	S 600	SK 605	XB 2809
Dynamit Nobel	PTMT		G 30	
Eastman	Tenite PTMT	6 PRO	6 H 91	
Hoechst	Hostadur B	VP 8600	VP 7600 GV 1/30	VP 7600 GV 1/30 SE-O
GAF	Gafite			
Hooker	(planned)			
Allied Chemical	Versel		1200	
Toray Industries				

properties. The glass temperature can be raised by introducing other diols into the polymer (see Ref. 3), and in view of the differences in the glass temperatures reported by the various companies, it is fully possible that some of the polymers given in Table 6-4 are not pure poly(butylene terephthalate), but rather are copolymers.

Poly(butylene terephthalates) have good stability to aliphatic hydrocarbons, alcohols, ether, perhalogenated hydrocarbons as well as to varying acids and bases. They are swollen by low-molecular-weight esters, ketones and partly halogenated hydrocarbons. They are soluble in warm *o*-chlorophenol, hexafluoroacetone hydrate and in mixtures of phenol/tetrachloroethane. However, no stress cracking accompanies this.

Mechanical properties are listed in Table 6-5. In general, there is good agreement among the various products. Only in the ultimate elongation is there much disagreement between the producers. The products show only a small tendency towards creep and can be used for short periods under loads at temperatures up to about 170° C. Their moisture absorption is quite small. Glass-fiber reinforced and flame retardant-containing varieties are also available. Property values for the 30% glass-fiber reinforced varieties are listed in Table 6-6.

Poly(butylene terephthalates) can be fabricated by injection molding and extrusion as well as by machining. Because of susceptibility to hydrolysis, the polymers must be thoroughly dry before processing. Fabrication temperatures for injection molding should be about 230–270° C, the equipment

Properties of Poly(tetramethylene terephthalates).

Property	Physical unit	Values measured for: Crastin S 600 Ref. 2	Hostadur B VP 8600 Ref. 7	PTMT Ref. 6	Tenite PTMT 6 PRO Ref. 5	Ultradur B 4500 Ref. 4
Density	g/cm^2	1.31	1.30	1.31	1.31	1.29
Melting temperature	°C	225	225	225	224	225
Glass temperature	°C	22[3]	36–39		43	
Heat distortion temperature (1.86 N/mm^2)	°C	50	65	65	50	67
Vicat softening temperature	°C		169	180		
Specific heat capacity (25 °C)	J g^{-1} K^{-1}		1.21		1.50	
ditto, for the melt			2.38		2.30	
Coefficient of linear thermal expansion	K^{-1}	8×10^{-5}	6.5×10^{-5}	7×10^{-5}	7.8×10^{-5}	
Tensile yield strength	N/mm^2	52	58	60	56	60
Yield at maximum elongation	%	5	6			
Ultimate elongation	%	250	40	200	250	200
Tensile strength	N/mm^2				35	
Modulus	N/mm^2	2700	2800	2600		2600
Flexural modulus	N/mm^2		2700		2300	
Flexural strength	N/mm^2	85	95	87	84	85
Hardness (indentation)	N/mm^2	126	130			150
(Rockwell R)	—		120			
(Rockwell M)	—		83		68	
(Shore D)	—	82				
Impact strength	—	No break	No break	No break	No break	No break
Impact strength with notch	J/m		28		53	
Dielectric constant	—	3.7	3.3	3.3	3.3	3.3
Volume resistivity	Ω cm	10^{15}	10^{16}	$> 10^{16}$	2×10^{15}	10^{16}
Surface resistivity	Ω	$> 10^{13}$	10^{13}	$> 10^{13}$		10^{13}
Dissipation factor	—	0.0013	0.0015	0.002	0.002	0.012

TABLE 6-6
Properties of Poly(tetramethylene terephthalates) reinforced with 30% glass fibers.

Property	Physical unit	Values measured for			
		Crastin Sk 605	Hostadur B 7600 GV2	PTMT G-30	Tenite PTMT 6H91
Density	g/cm^3	1.53	1.52	1.53	1.53
Melting temperature	°C	225	220–225	225	
Glass temperature	°C		36–49		
Heat distortion temperature ($1.86\ N/mm^2$)	°C	205	208	>200	207
Vicat softening temperature	°C		216	200	
Coefficient of linear thermal expansion	K^{-1}	3×10^{-5}	$(4\text{–}5) \times 10^{-5}$	4.5×10^{-5}	
Tensile yield strength	N/mm^2	133		60	
Tensile strength	N/mm^2		150	140	130
Ultimate elongation	%	3.5	4	<2	5
Modulus	N/mm^2	8500	10,000	10,000	8000
Flexural strength	N/mm^2	202		190	197
Flexural modulus	N/mm^2		8500		8400
Impact strength	J	no break	400	380	750
Notched impact strength	J/m	100	90	110	86
Hardness (identation)	N/mm^2	176	195	173	
(Rockwell R)	—		120		
(Rockwell M)	—		94		90
(Shore D)	—		86		
Dielectric constant	—	3.8	3.8	3.7	
Dissipation factor	—	0.0015	0.0020	0.0025	
Surface resistivity	Ω	$>10^{15}$	10^{13}	$>10^{13}$	
Volume resistivity	Ω cm	$>10^{15}$	10^{16}	$>10^{16}$	
Dielectric strength	kV/mm	20	60	29	29
Arc resistance	s	110			
Water absorption	%	0.15	0.05		
LOI	%				18.5

temperatures about 30–80°C. Thanks to rapid crystallization, short cycle times are possible. The polymer begins to decompose over 270°C with discoloration and with a drop in the values of the mechanical properties. Heat zones must be well regulated during extrusion.

Poly(butylene terephthalates) are used for bearing bushings, gear wheels, rollers, drums, pump parts and in the building of computers and measuring instruments and in industry. Films are used in the food and cosmetic industries as decorative and insulating materials because of their low permeability.

References

1. H. Pfister and H. Bopp, *Kunststoff-Technik*, **11**, 307 (1973).
2. Product information literature of Ciba Geigy; F. Breitenfellner, *Kunststoffe*, **65**, 743 (1975).
3. J. Habermeier, L. Buxbaum and H. Batzer, *Chem.-Ztg.*, **98**, 184 (1974).
4. Product information literature of BASF.
5. Product information literature of Eastman Kodak.
6. Product information literature of Dynamit Nobel.
7. Product information literature of Hoechst.

6.5 POLY(1, 4-CYCLOHEXYLENEDIMETHYLENE TEREPHTHALATE-CO-ISOPHTHALATE)

This copolymer contains 1,4-cyclohexanedimethylol units (I), terephthalic acid units (II) and isophthalic acid units (III)

$-OCH_2-$ ⟨cyclohexylene⟩ $-CH_2O-$ (I) $-OC-$ ⟨p-phenylene⟩ $-CO-$ (II) $-OC-$ ⟨m-phenylene⟩ $-CO-$ (III)

Neither the mole ratio of II and III, nor the ratio of *cis*- and *trans*- structures in I are published. The copolymer is marketed by Eastman Kodak under the tradenames Tenite Polyterephthalate PCDT®, Tenite Polyterephthalate 7 DRO®, or Kodar®.

In contrast to the poly(tetramethylene terephthalate) varieties, the copolymer is amorphous and transparent and gives molded articles with excellent hardness and tensile strength (Table 6-7). In addition it can be extruded to give films. Films and molded articles can be used in food packaging without reservation. The films can be sterilized with ethylene oxide. Kodar® competes with PVC in its application for deep-drawn blister packages which allow higher filling temperatures.

References

1. Product information literature of Eastman Kodak.

6.6 POLY(BUTYLENE TEREPHTHALATE-CO-(MULTIBUTYLENEOXY) TEREPHTHALATE)

6.6.1 Structure and Synthesis

These polymers, also known as polyether–ester block copolymers, are copolymers of "soft" segments out of multibutyleneoxy terephthalate blocks

TABLE 6-7
Properties of copolymers of terephthalic acid, isophthalic acid and cyclohexane-1,4-dimethylol (Tenite 7 DRO).

Property	Physical unit	Values measured for: Molded article	Extruded films: 25.4 μm	Extruded films: 127 μm
Density	g/cm^3	1.2		
Transparency	%		80	70
Gloss at 45°	Gardner Scale		110	110
Haze	%		0.4	0.4
Coefficient of linear thermal expansion	K^{-1}	7.9×10^{-5}		
Heat distortion temp. (at 1.86 N/mm^2)	°C	68		
Tensile yield strength	N/mm^2	51.3	49	49
Tensile strength	N/mm^2		39	39
Ultimate elongation	%	210	25	100
Modulus	N/mm^2		1800	1800
Flexural modulus	N/mm^2	2100		
Flexural strength	N/mm^2	75		
Compressive modulus	N/mm^2	1230		
Impact strength		no break		
Impact strength with notch	J/m	53		
Deformation under 7 N/mm^2	%	0.7		
Rockwell Hardness (R)	—	108		
Water vapor permeability	$g\ cm^{-2}\ s^{-1}$	—	5.4×10^{-8}	1.08×10^{-8}
Gas permeability, CO_2	$cm^3\ s\ g^{-1}$	—	2.9×10^{-15}	2.9×10^{-15}
O_2	$cm^3\ s\ g^{-1}$	—	1.4×10^{-15}	1.4×10^{-15}
N_2	$cm^3\ s\ g^{-1}$	—	0.7×10^{-15}	0.7×10^{-15}

(I) I $\left[O(CH_2CH_2CH_2CH_2O)_n-OC-C_6H_4-CO \right]_x$ $(PTMEG\text{-}T)_x$

with "hard" segments containing butylene terephthalate or tetramethylene terephthalate units (II)

II $\left[OCH_2CH_2CH_2CH_2O-OC-C_6H_4-CO \right]_y$ $(4GT)_y$

These types of polymers are marketed by DuPont as thermoplastic elastomers under the tradename Hytrel®. The use of similar polymers as elastic fibers,[1-11] laminates,[12] adhesives,[13] and as modifiers for poly-(ethylene terephthalate) plastics[14] has also been described.

Synthesis of the polymer goes back to work in the 50's,[1,2] which led to the synthesis of homo- and copolymers with tetramethylene terephthalate units.[15-26] It follows the classical poly(ethylene terephthalate) synthesis, that is, transesterification of dimethyl terephthalate with 1,4-butanediol or with a mixture of 1,4-butanediol and poly(butylene oxide),

$$2\ CH_3OOC-C_6H_4-COOCH_3 + 3\ HO-G-OH \xrightleftharpoons{200\ °C} H(O-G-O-OC-C_6H_4-CO)_2O-G-OH + 4\ CH_3OH \qquad (6\text{-}5)$$

where $G=(CH_2)_4$ or $(CH_2)_4(O(CH_2)_4)_m$, followed by a polycondensation in the second step

$$2\ -OC-C_6H_4-CO-O-G-OH \rightleftharpoons -OC-C_6H_4-CO-O-G-O-OC-C_6H_4-CO- + HO-G-OH \qquad (6\text{-}6)$$

The resulting polyether–esters have molecular weights ranging between 25,000 and 30,000 g/mol. They have hard segments $(4GT)_y$ with high glass temperatures and soft segments $(PTMEG\text{-}T)_x$ with lower glass temperatures. In a copolymer having 57.5 wt.-% (87.4 mol %) 4GT-units and 42.5 wt.-% PO4T-units (molecular weight of the polyether, at about 1000 g/mol), at least 36% of the rigid segments contain ten or more 4GT-units.

Out of the many possible polyether constituents, only poly(tetrahydrofurane) is suitable, for only with it is the full hardness of the polymer obtained directly after fabricating. In contrast, polyether–esters with poly(ethylene oxide) soft segments require 24–36 h to reach full hardness after fabricating.

6.6.2 Properties, Fabrications and Application

According to electron microscope pictures, polyether–esters are composed of two phases.[21] The combination of low glass temperature and high melting point (Table 6-8) makes them thermoplastic elastomers, that is, polymers which may be processed like thermoplastics, but which have elastomeric properties. The vulcanization step is omitted altogether. Use temperatures fall between −55 and +150°C. The polyether–esters show high resiliency and a good split-tear and abrasion resistance.

The Hytrel polymers are soluble in *m*-cresol; those with less than 60% hard segments are also soluble in methylene chloride, chloroform and 1,1,1-trichloroethane.[22] They are resistant to oils and many chemicals. Hytrel polymers have a low melt-viscosity. As a result of high crystallization rates of the hard segments, short injection molding cycle times and high extrusion rates are possible. Thermal post-treatment is not necessary. The

TABLE 6-8
Properties of Hytrel–polyether esters.[20,22,24,26]

Property	Physical unit	Values measured for differing percentages of 4GT-rigid segments 33	58	76
Density	g/cm^3	1.15	1.20	1.22
Melting temperature	°C	176	202	212
Glass temperature	°C	−78	−50	−2
Brittleness temperature	°C	< −70	< −70	< −70
Tensile strength	N/mm^2	40	45	49
Ultimate elongation	%	810	760	510
Modulus 5%	N/mm^2	2.2		
10%	N/mm^2	3.7	10.0	17.2
100%	N/mm^2	8.1	15.1	20.1
300%	N/mm^2	11.0		
Flexural modulus	N/mm^2	45.8	2110	507
Notched impact strength	N	> 1090	> 1090	> 1090
Falling-ball rebound	%	700		
Shore Hardness	—	A 92	D 55	D 63
Abrasion resistance	%	700		
Tear	N/mm	294	353	
Split tear	N/mm	16		
Compression set (95 N/mm^2 at 70 °C)	%		4	3

polymers can be fabricated by rotational molding, melt casting, fluidized-bed sintering and electrostatic spraying.

Typical applications are water hoses, low pressure tires, transmission belts, cable sheathings, and drive belts for snow mobiles.

References

1. U.S. Patent 3023192 (May 29, 1958); DuPont; inv.: J. C. Shivers, Jr.; *C. A.*, **56**, 13071f (1962).
2. W. H. Charch and J. C. Shivers, *Textile Res. J.*, **29**, 536 (1959).
3. French Patent 1354553 (March 6, 1964); DuPont; inv.: P. E. Frankenburg; *C. A.*, **62**, 85423h (1965).
4. U.S. Patent 3157619 (November 10, 1961); Eastman Kodak; inv.: A. Bell, C. J. Kibler and J. G. Smith; *C. A.*, **62**, 2904c (1965).
5. U.S. Patent 3238178 (October 16, 1961, January 15, April 9 and August 9, 1962); Eastman Kodak; inv.: C. J. Kibler, A. Bell and J. G. Smith; *C. A.*, **62**, 4195d (1965).
6. U.S. Patent 3243413 (October 18, 1962); Eastman Kodak; inv.: A. Bell, C. J. Kibler and J. G. Smith; *C. A.*, **65**, 4079b (1966).

7. U.S. Patent 3261812 (January 15, 1962); Eastman Kodak; inv.: A. Bell, C. J. Kibler and J. G. Smith; *C. A.*, **62**, 4195d (1965).
8. U.S. Patent 3277060 (January 15, 1962); Eastman Kodak; inv.: A. Bell, C. J. Kibler and J. G. Smith; *C. A.*, **62**, 4195d (1965).
9. A. A. Nishimura and H. Komagata, *J. Macromol. Sci. [Chem.]* **A1**, 617 (1967).
10. Japanese Patent Appl. 44-20469/1969 (1964); Asahi Kasei Kogyo; inv.: H. Kobayashi and K. Sasaguri.
11. E. Leibnitz and G. Reinisch, *Faserforschg. Textiltechnik*, **21**, 426 (1970).
12. U.S. Patent 2865891 (December 23, 1958); DuPont; inv.: R. H. Michel; *C. A.*, **53**, 23084 (1959).
13. U.S. Patent 3013914 (December 19, 1961); DuPont; inv.: A. Willard; *C. A.*, **56**, 11822i (1962).
14. French Patent 2012585 (1969); Hoechst; *C. A.*, **73**, 99711z (1970); Brit. Pat. 1247759; Hoechst; inv.: H. Fröhlich and L. Brinkmann; *C. A.*, **73**, 99711z (1970).
15. U.S. Patent 3651014; German Patent 2035333; DuPont; inv.: W. K. Witsiepe; *C. A.*, **74**, 143 109f (1971).
16. U.S. Patent 3763109; German Patent 2240801 (March 1, 1973); DuPont; inv.: W. Witsiepe; *C.A.*, **79**, 19 440w (1973).
17. U.S. Patent 3766146; German Patent 2213128 (September 21, 1972); DuPont; inv.: W. K. Witsiepe; *C. A.*, **78**, 17337y (1973).
18. U.S. Patent 3784520 (1974); DuPont; inv.: G. K. Hoeschele.
19. M. Brown and W. K. Witsiepe, *Rubber Age*, **104**(3), 35 (1972).
20. W. K. Witsiepe, Segmented polyester thermoplastic elastomers, in N. J. Platzer (ed.), *Adv. Chem. Ser.*, **129**, 39 (1973).
21. R. J. Cella, *J. Polymer Sci. [Symp.]*, **42**, 727 (1973).
22. G. K. Hoeschele and W. K. Witsiepe, *Angew. Makromol. Chem.*, **29/30**, 267 (1973).
23. W. H. Buck and R. J. Cella, *ACS Polymer Preprints*, **14**(1), 98 (1973).
24. G. K. Hoeschele, *Chimia [Aarau]*, **28**, 544 (1974).
25. W. H. Buck, R. J. Cella, Jr., E. K. Gladding and J. R. Wolfe, Jr., *J. Polymer Sci. [Symposia]*, **48**, 47 (1974).
26. G. K. Hoeschele, *Polymer Engng. Sci.*, **14**, 848 (1974).

6.7 SATURATED POLYTEREPHTHALATES

The firm Dynamit Nobel produces for a variety of applications a series of saturated copolyesters out of terephthalic acid (or its dimethyl ester), other aromatic or aliphatic dicarboxylic acids and diols under the tradenames Dynapol or Polyester (see Table 6-9). These copolymers are either linear, if they are made with diols, or branched, if they are made with triols (or higher alcohols). Therefore, their properties can vary considerably. The number of special varieties which are available makes it necessary to leave out detailed description of their individual properties.

References

1. Product information of Dynamit Nobel.

TABLE 6-9
Names and designations as well as areas of application of saturated copolyesters of the firm Dynamit Nobel AG.

Name and type	Structure	Combined with	Area of application
Dynapol L 205 L 206	linear	certain melamine or benzoguanamine resins	boilable, sterilizable, and highly deformable coating finishes
Dynapol L 411	linear	certain melamine or benzoguanamine resins	deep drawable and stampable sheet and strip coatings for metals
Polyester L 1850	linear	certain melamine or benzoguanamine resins	deep drawable and stampable sheet and strip coatings for metals
Dynapol P 300	branched, cross-linkable, hydroxyl-group containing	suitable cross-linkers, *i.e.* blocked isocyanates	electrostatic powder coating of metals
Dynapol H 700	branched, cross-linkable, hydroxyl-group containing	suitable cross-linkers, *i.e.* blocked isocyanates	for coil coating and conventional painting
Dynapol H 702	branched, cross-linkable, hydroxyl-group containing	suitable cross-linkers, *i.e.* blocked isocyanates	for coil coating
Polyester, S-Series	?	—	melt adhesive with glass temperatures between −17 and +30°C
Polyester P 15000	?	—	binder for fluidized bed sintering
Polyester LP-C 18811/3S	linear	—	coil coating
Hot melt TS HM or HME	?	—	melt adhesive for the shoe industry

6.8 POLYARYLATES

Polyarylates are defined as polyesters from diphenols and dicarboxylic acids (for a review see Ref. 1). The first commercial polyarylate from terephthalic acid (I), isophthalic acid (II), and bisphenol A (III) in the molar ratio 1:1:2 is manufactured by Unitika Ltd.[2,3] since 1974 under the name U-polymer[2]

HOOC—C_6H_4—COOH (I) HOOC—C_6H_4—COOH (II) HO—C_6H_4—$C(CH_3)_2$—C_6H_4—OH (III)

The synthetic route was not disclosed. Development work is also carried out by Dynamit Nobel.[1]

The U-polymer is a light brown, transparent, X-ray amorphous material with a density of 1.21 g/cm^3 and a refractive index of 1.61. The amorphous character does not change on annealing. The glass transition temperature is given as 173°C[3] whereas another reference[1] gives 194°C for a polymer with the same composition. The heat deflection temperature is 164°C (at 1.84 N/mm^2).[3] The coefficient of thermal expansion (6.2×10^{-5} K^{-1}) is a little bit smaller than that of polycarbonates. The polymer is self-extinguishing.

The mechanical properties are those of a typical engineering plastic: tensile strength of 74 N/mm^2, tensile modulus of 2200 N/mm^2, elongation at break of 62 %, flexural strength of 93 N/mm^2, flexural recovery of only 2–3 %, compressive strength of 94 N/mm^2, compressive modulus of 21 N/mm^2, and Rockwell hardness (M scale) of 93. The mechanical properties at elevated temperatures compared also very well with those of other engineering plastics. The impact strength of polyarylate exceeds all other plastics except polycarbonate.[3]

The electrical properties are also good: volume resistivity of 2×10^{-16} Ω cm, arc resistance of 129 s, relative permittivity of 3 (at 1 MHz), and dielectric loss factor of 0.015 (at 1 MHz).

The U-polymers can be injection molded into electrical, automotive, and mechanical parts and into domestic appliances. Blow molding leads to bottles and vessels. Extrusion gives pipes, rods, sheets, films and monofilaments.

References

1. G. Bier, *Polymer [London]*, **15**, 527 (1974).
2. K. Hazama, *Jap Plastics*, **8**(3), 6 (1974); *ibid.*, *Japan Plastics Age*, Oct. 1974.
3. H. Sakata, *SPE Ann. Techn. Papers*, **20**, 459 (1974).

7 Sulfur Polymers

7.1 POLY(PHENYLENE SULPHIDE)

7.1.1 Structure

Poly(phenylene sulfide) is the trivial name for poly(thio-1,4-phenylene) with the base unit

$$-\!\!\left[S\!-\!C_6H_4\right]\!\!-$$

The abbreviation PPS is also frequently used. Commercial products can be either linear or branched. A nearly linear product was offered as a trial product by The Dow Chemical Company under the name Experimental Resin QX-4375.1® in 1964.[1] Phillips Petroleum has produced linear and branched products under the tradename Ryton® since 1973.[2,3]

7.1.2 Synthesis

The first poly(phenylene sulfides) were probably made as early as 1897 by the Friedel-Crafts reaction of sulfur with benzene:[4]

$$C_6H_6 + S \xrightarrow{AlCl_3} \left[C_6H_4-S \right] + \text{thianthrene} \qquad (7\text{-}1)$$

Even in the absence of aluminum trichloride the thermal polycondensation of benzene and sulfur leads to resin-like substances.[5] Other early attempts started out from thiophenol either in the presence of aluminum trichloride,[6] or in concentrated sulfuric acid[7] or in the presence of thionyl chloride.[8] All of these reactions gave amorphous, resinous products with the composition C_6H_4S, which decomposed between 200–300°C. Similar products were obtained from the reaction of phenol with sulfur dichloride.[9,10]

The so-called Macallum synthesis aroused greater technical interest:[11-15]

$$Cl-C_6H_4-Cl + S + Na_2CO_3 \xrightarrow{275-360\,^{\circ}C} \left[C_6H_4-S_n \right] \qquad (7\text{-}2)$$

The Macallum synthesis is not reproducible,[14,15] and invariably gives branched and cross-linked products, which generally have $n = 1.0–1.3$ sulfur atoms per base unit. The reaction is very exothermic and even on a laboratory scale, is difficult to control.[16,17] The Macallum patents were sold to The Dow Chemical Company in 1954. Research efforts at the company initiated at that time led to a new synthesis, namely the self-condensation of the metal salts of *p*-halothiophenol:[15,18]

$$X{-}C_6H_4{-}SMt \xrightarrow{200-250\,^{\circ}C} {-}[C_6H_4{-}S]{-} + MtX \qquad (7\text{-}3)$$

X can be F, Cl, Br, I or $N_2^+X^-$; Mt can be, for instance, Cu, Li, Na or K. The reaction is carried out in solvents such as pyridine. However, byproducts from this synthesis are difficult to remove, which presumably led to termination of the experimental production. A review article of these now historic syntheses was published in 1969.[1]

In comparison, the commercial synthesis used by Phillips Petroleum Company starts with dichlorobenzene and sodium sulfide;[3,19] the reaction is apparently carried out in *N*-methylpyrrolidone:

$$Cl{-}C_6H_4{-}Cl + Na_2S \xrightarrow{\Delta} {-}[S{-}C_6H_4]{-} + 2\,NaCl \qquad (7\text{-}4)$$

The polymer is cured by heating it under oxygen whereby a number of processes take place.[27]

7.1.3 Properties

PPS is a fine white powder,[9] which discolors on heating in air to a brown color.[3,20] This apparently is the result of cross-linking, since the brown product is insoluble in all solvents.[3] Presumably then, the white product is linear.[21] Molecular weights are not known exactly, but they apparently are not too high, since the inherent viscosity is only about 16 ml/g in 1-chloronaphthalene at 206°C.[3]

The white Phillips polymer is highly crystalline.[22] The crystal structure appears similar to that of poly(oxy-1,4-phenylene). The melting temperature of one of the commercial polymers is given as 288°C,[20] that of an annealed polymer reported to be weakly cross-linked as 295°C.[22] The glass temperature is not known exactly: a value of 85°C is given for the Phillips polymer,[3] in contrast to the value of 150°C given for the Dow experimental resin.[1] The Vicat softening temperature is 315°C.[20] Neither in air nor under nitrogen was there any noticeable weight-loss under 500°C according to thermogravimetric measurements.[3,20] There is nearly total decomposition at 700°C in air (Figure 7-1). About 40% of the mass remains intact on heating under nitrogen to 1000°C with the rate of heating at 10 K/min.[3] The thermal stability of the Phillips polymers is supposed to be 40–50°C higher than that of the Dow experimental resin.[3] The recommended maximum use-temperature is not to exceed 260°C.[3]

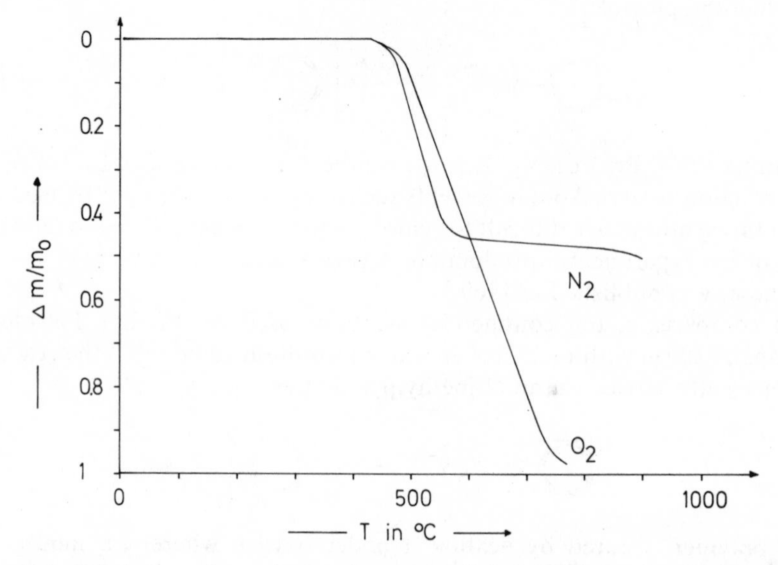

Figure 7-1 Relative loss of mass $\Delta m/m_0$ of an uncross-linked PPS during thermal degradation under nitrogen (N_2) or in air (O_2). Rate of heating is 10 K min^{-1} (data from Ref. 26).

The uncross-linked Phillips polymers are very stable to organic and inorganic acids, aqueous bases, certain amines, hydrocarbons, organic esters, most halogenated hydrocarbons, aldehydes, ketones, nitriles, nitrobenzene, alcohols, phenols, aqueous solutions of inorganic salts and a number of other organic solvents. This stability was determined by measuring the tensile strengths after 24 h at 93°C in these solvents.[24] PPS is attacked by formic acid, certain amines, a few halogenated hydrocarbons, benzaldehyde, nitromethane, certain ethers, aqueous solutions of sodium hypochlorite and bromine water. At higher temperatures PPS is soluble in, for instance, 1-chloronaphthalene. PPS cross-linked in air is insoluble in all these solvents. Concentrated sulfuric acid decomposes PPS rapidly.[25]

Figure 7-2 shows a curve of the specific heat capacity as a function of temperature.[26] The thermal conductivity is about 0.30 W m^{-1} K^{-1} between 40 and 167°C.[26] Figure 7-3 reproduces the temperature dependence of the coefficient of linear thermal expansion.[26]

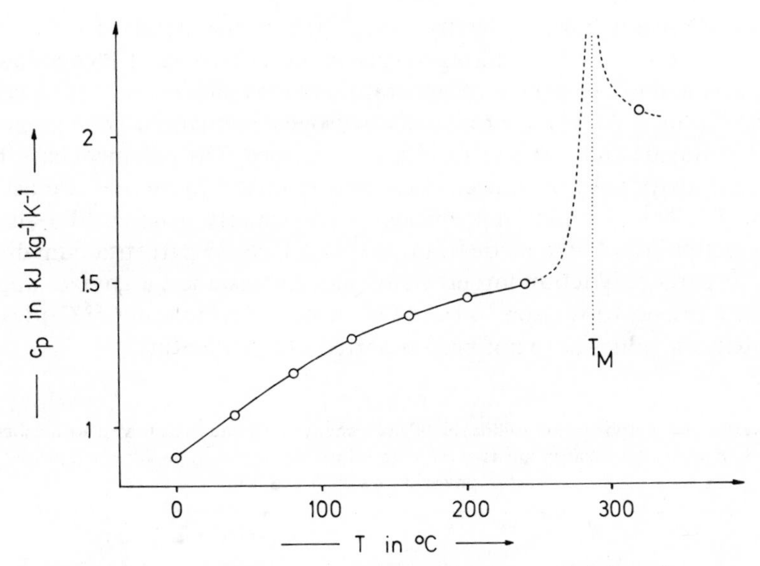

Figure 7-2 Specific heat capacity c_p at constant pressure as a function of temperature T (according to data in Ref. 26). T_M=melting temperature.

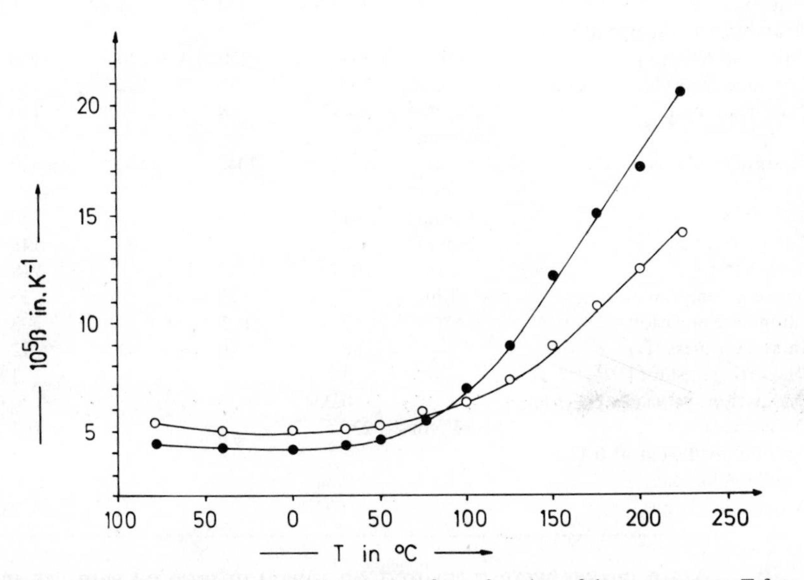

Figure 7-3 Coefficient of linear thermal expansion β as a function of the temperature T for an uncross-linked (●) and a cross-linked (○) PPS (data from Ref. 26). The glass temperature, obtained by extrapolating the linear portions of the curves to their intercept, is $T_G \approx 85\,°C$.

PPS does not burn under the conditions of the standard tests ASTM D-635 and UL 94.[26] The limiting oxygen index (LOI) is 44.[3] PPS does burn as long as a flame is present, with generation of gray smoke and with a yellow-orange color.[26] At 815°C principally hydrogen, methane, carbon monoxide, carbon dioxide and carbon oxysulfide are formed. The polymer chars, but it does not drip.[23] Typical mechanical and electrical properties are listed in Table 7-1. The low friction coefficient is particularly striking. PPS also has poor wettability. A film made from 100 parts PPS, 33 parts titanium dioxide and 10 parts poly(tetrafluoroethylene) has, for instance, a contact angle of 68° with respect to Wesson® oil and 110° with respect to water.[25] Critical surface tension values have not been reported in the literature.

TABLE 7-1
Properties of poly(phenylene sulfide) or of poly(phenylene sulfide) reinforced with asbestos or glass fibers.[3,21] All measurements were determined on injection molded samples at 21 °C unless otherwise noted.

		Values measured for			
Property	Physical unit	PPS 100	PPS/ Asbestos 80/20	PPS/ Asbestos 50/50	PPS/Glass 60/40
Density	g/cm^3	1.35	1.52	1.67	1.65
Heat distortion temperature (at 1.86 N/mm^2)	°C	138	>220	>220	>220
Continuous service temperature	°C	260		260	260
Tensile strength, 21 °C	N/mm^2	64–77	84	76	150
204 °C	N/mm^2	33			33
Flexural modulus 21 °C	N/mm^2	4200	7040	8400	15,500
204 °C	N/mm^2				4220
Flexural strength	N/mm^2	141	155		250
Compressive strength	N/mm^2	113		169	148
Impact strength	J/m	109	163		218
Impact strength with notch	J/m	22	22		55
Ultimate elongation	%	3	2		2–3
Shore Hardness (D)	—	86	89	91	92
Dielectric constant (10^3–10^6 Hz)	—	3.1			3.8
Dissipation factor (1 kHz)	—	0.004			0.0037
Dielectric strength	kV/cm	230			193
Friction coefficient at 0 U/min and 6.8 kg load	%	0.64			0.50
Water absorption	%	<0.02			<0.05

Stress/strain curves were measured on injection molded samples in the temperature range of −40 to +260°C.[26] As expected, glass-fiber reinforced polymer shows higher tensile strength at maximum load at temperatures under

150°C (Figure 7-4). The elongation of filled PPS remains practically unchanged between −40 and +260°C, which apparently reflects the influence of the glass fibers. On the other hand, the elongation of unfilled polymer increased sharply from about 2 to 25% between 65 and 93°C (not shown), followed first by an additional increase to 100% at 205°C and then by a drop to 2% at 260°C. Presumably, this unusual behavior is attributed to the influence of the glass temperature and the melting temperature.

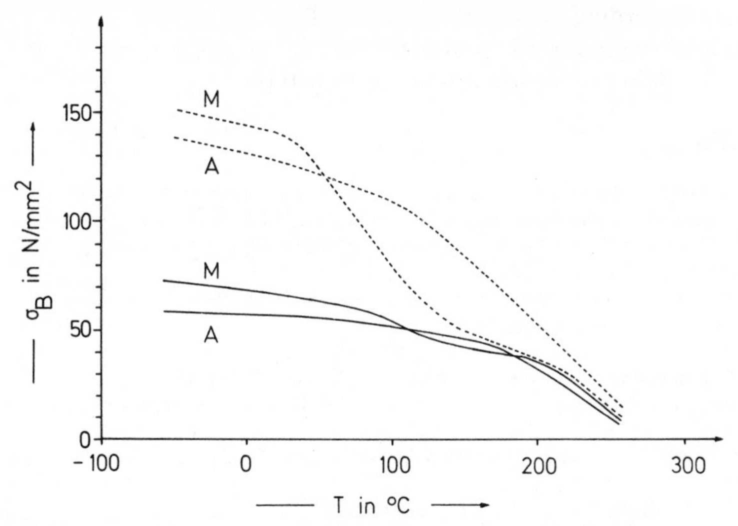

Figure 7-4 Tensile strength σ_B as a function of the temperature for cross-linked (– – –) and uncross-linked (———) PPS. M = measurements directly on the injection-molded sample, A = measurements after annealing (data from Ref. 26).

7.1.4 Fabrication

Filled or unfilled PPS can be cast, injection molded and extruded. Injection molding should be carried out at temperatures of 310–375°C with equipment at 65 to 150°C.[21] On compression molding the mold is filled with PPS, subjected to pressures of 140–210 kg/cm^2 in the cold, and then heated to at least 315°C. Finally it is subjected to pressures of 70–280 kg/cm^2. Cooling must be controlled with a maximum cooling rate of about 2 K/min until about 230°C is reached, otherwise cracks and holes are formed. The part can be removed from the mold at about 150°C. Both the filled and the unfilled polymers must be "cross-linked" through air oxidation at elevated temperatures before fabrication in order to increase the melt viscosity. These reaction times amount to between one hour and several hours at temperatures up to about 370°C depending on the density of the material. Coatings can be applied also through fluidized bed sintering or flame spraying.

7.1.5 Application

PPS has good thermal stability, excellent chemical stability, a low friction coefficient, good abrasion resistance and very good electrical properties. This exceptional engineering material is used, therefore, for coatings on glass, aluminum, titanium, steel and bronze as well as for molded parts. Examples are valves, pump parts and bearing boxes for grease-free ball bearings. Other applications include coatings for cooking pots and pans. These coatings are not toxic according to animal feeding studies.

Production amounted to about 3000 t/a, the price about $7.50 $/kg in 1973. The 1976 production was about 25,000 t/a.

References

1. H. A. Smith, Poly(arylene sulfides), *Encycl. Polymer Sci. Technol.*, **10**, 653 (1969).
2. Anonym., *Chem. Eng. News*, **49**(29), 17, 44 (July 19, 1971).
3. J. N. Short and H. W. Hill, Jr., *Chem. Tech.*, **2**, 481 (1972).
4. P. Genvresse, *Bull. Soc. Chim. France* [*3*], **17**, 599 (1897).
5. H. B. Glass and E. E. Reid, *J. Amer. Chem. Soc.*, **51**, 3428 (1928).
6. J. J. Deuss, *Rec. Trav. Chim.*, **28**, 136 (1909).
7. T. P. Hilditch, *J. Chem. Soc.* [*London*], **97**, 2579 (1910).
8. H. S. Tasker and H. O. Jones, *J. Chem. Soc.* [*London*], **131**, 1141 (1928).
9. E. Ellis, *The Chemistry of Synthetic Resins*, Vol. II, Reinhold, New York, pp. 1183, 1189 (1935).
10. N. J. Gaylord, *Polyethers* (High Polymers Series XIII, Part 3), Interscience, New York, p. 31 (1962).
11. A. D. Macallum, *J. Org. Chem.*, **13**, 154 (1948).
12. U.S. Patent 2513188 (June 27, 1950); inv.: A. D. Macallum; *C. A.*, pp. 44–8165a (1950).
13. U.S. Patent 2538941 (January 23, 1951); inv.: A. D. Macallum; *C. A.*, pp. 45–5193c (1951).
14. R. W. Lenz and W. K. Carrington, *J. Polymer Sci.*, **41**, 333 (1959).
15. R. W. Lenz and C. E. Handlovits, *J. Polymer Sci.*, **43**, 167 (1960).
16. A. D. Macallum, cited in Ref. 3.
17. H. A. Smith and C. E. Handlovits, ASD-TDR-62-372, Report on the Conference on High Temperature Polymer and Fluid Research, Dayton, Ohio, p. 100, cited in Ref. 3.
18. R. W. Lenz, C. E. Handlovits and H. A. Smith, *J. Polymer Sci.*, **58**, 351 (1962).
19. U.S. Patent 3354129 (November 21, 1967); Phillips Petroleum Co.; inv.: J. T. Edmonds, Jr. and H. W. Hill, Jr.; *C. A.*, **68**, 13598e (1968).
20. H. W. Hill, Jr. and E. T. Edmonds, Jr., *Amer. Chem. Soc., Div. Org. Coatings Plastics Chem., Preprints*, **30**(2), 199 (1970).
21. H. W. Hill, Jr., R. T. Werkman and G. E. Carrow, *Amer. Chem. Soc., Div. Org. Coatings Plastics Chem., Preprints*, **33**(2), 180 (1973); *Adv. Chem. Ser.*, **134**, 149 (1974).
22. B. J. Tabor, E. P. Magré and J. Boon, *Europ. Polymer J.*, **7**, 1127 (1971).
23. S. Tsunawaki and C. C. Price, *J. Polymer Sci.*, **A2**, 1511 (1964).
24. D. G. Brady and H. W. Hill, Jr., *Modern Plastics* (May 1974) p. 60.
25. H. W. Hill, Jr. and J. T. Edmonds, Jr., *Amer. Chem. Soc., Polymer Preprints*, **13**, 603 (1972); *Adv. Chem. Ser.*, **129**, 80 (1973).
26. Phillips Petroleum Co., Ryton® PPS—Physical and chemical properties, TSM 266, July 1974.
27. R. T. Hawkins, *Macromolecules*, **9**, 189 (1976).

7.2 ALIPHATIC POLYSULFIDES.

7.2.1 Structure

Aliphatic polysulfides have either the base unit I or the structural element II

$$\text{I} \quad -\!\!\left(S_x - R\right)\!\!- \qquad\qquad \text{II} \quad -\!\!\left(S - R - S - R'\right)\!\!-$$

Aliphatic polysulfides like type I with varying sulfur grades have been produced by the Thiokol Corp. for many years. Aliphatic polysulfides with the structural element II result from curing radiation curable polymers or photopolymer systems recently introduced by W. R. Grace and Co. These compounds are often designated with the abbreviation RCP (= radiation curable polymers).

7.2.2 Prepolymers

The RCPs are low or high molecular mixtures of polythiols and polyenes with at least two functional groups per molecules. The structures of these compounds can vary widely; they have not been disclosed by the manufacturer. W. R. Grace and Co. currently offers about 15 varieties of prepolymers, which all differ in viscosity, surface tension, curing rates etc. The consistencies can range from low-viscosity fluids to grease-like substances. They are stable providing that they are not exposed to light, temperatures over 60°C or radical sources. Prepolymers should be handled only under yellow, fluorescent light. The shelf life is about 6 months. All the RCPs can be mixed with each other. Dyes and pigments are compatible with the RCPs.[1]

7.2.3 Curing

The curing reaction consists of a radically induced addition of thio groups to the double bonds of the polyenes (see Ref. 2):

$$\sim\!\!\sim R{-}SH + CH_2{=}CH\sim\!\!\sim \longrightarrow \sim\!\!\sim R{-}S{-}CH_2{-}CH_2\sim\!\!\sim \qquad (7\text{-}5)$$

For instance, peroxides, electron beam or ultraviolet light activated benzophenone can serve as the radical source. Curing with ultraviolet light is generally the most economical method. Wavelengths should be between 300–400 nm with the maximum as close as possible to 365 nm. The curing occurs in seconds or minutes according to the prepolymer type, the thickness of the layer etc. It is not measurably influenced by oxygen in air. UV irradiation permits curing of layer thicknesses of about 1 cm. Because either the polyene or the polythiols are multifunctional, cross-linked polymers result from the curing. The same process is used in the W. R. Grace Letterflex process for relief plate manufacture.

7.2.4 Properties

The properties of the cured polymers depend strongly on the properties of the prepolymer (Table 7-2). A few types are light stable, others yellow readily. They may be elastic or hard and stiff. Several cured polymers are very abrasion resistant; others, on the other hand, are flame resistant. The chemical, physical and toxicological properties of the polymers have not yet been fully investigated.

TABLE 7-2
Properties of several polythiol/polyene systems at 25 °C.[1]

Property	Physical unit	Value measured for type			
		411-D	611-U	2311-A	3861-C
Prepolymer					
Color	APHA	80	80	500-yellow	140
Viscosity	Pa s	0.050	0.40	0.664	274
Surface tension	N/cm	3.86×10^{-4}	4.0×10^{-4}	3.96×10^{-4}	
Density	g/cm^3	1.191	1.231	1.400	1.116
Cured products					
Tensile stress	N/mm^2	1	20	16.2	17.6
Ultimate elongation	%	30	35	122	117
Modulus	N/mm^2	5.4	1120	873	72
Shore hardness, A	—	70	90	69	80
D	—	26	81	50	46
Density	g/cm^3	1.291	1.29	1.46	1.18
Water vapor permeability	g/(cm s)		6.7×10^{11}	1.3×10^{11}	4.7×10^{11}
Water absorption	%		0.19	0.08	1.48
LOI	%		26	26.5	19.2
Dielectric constant (1 MHz)	—		4.04	3.92	
Dissipation factor (1 MHz)	—		0.045	0.034	
Volume resistivity	Ω cm		10^{15}	7×10^{14}	
Surface resistivity	Ω		10^{16}	3×10^{16}	
Arc resistance	s		95	3.2	
Dielectric strength	kV/cm		386	394	

7.2.5 Application

The RCPs are solvent-free systems and consequently are suitable as coatings for a series of applications in which high gloss, favorable burning properties, good dielectrical properties, good abrasion resistance *etc.* are necessary. They are suitable, also, as viscosity improvers, adhesive agents or as encapsulating agents for electronic components.

References

1. W. R. Grace and Co., Technical Literature, 1974.
2. C. S. Marvel and R. R. Chambers, *J. Amer. Chem. Soc.*, **70**, 993 (1948).

7.3 POLYSULFIDE ETHERS

7.3.1 Structure

Polysulfide ethers contain both the thio group —S—as well as the ether group —O— in the main chain. The commercially produced polymers are all derivatives of poly(thio-1,4-phenylenoxy-polymethylene)

$$-\!\!\left[S-C_6H_4-O-(CH_2)_x \right]\!\!- \qquad x = 4 \text{ or } 5$$

The Dow Chemical Company produces a few monomers in experimental quantities (see below).

7.3.2 Synthesis

The starting monomers can be made by three different routes. Phenols, for instance, are treated with a 10–100% excess of tetrahydrothiophene-1-oxide:[1-9]

$$C_4H_8S{=}O + C_6H_5OH + HCl \xrightarrow[0-10\,^{\circ}C]{H_2O,\ ROH} C_4H_8\overset{\oplus}{S}-C_6H_4-OH + Cl^{\ominus} + H_2O \qquad (7\text{-}6)$$

Alternatively, tetrahydrothiophene itself can be used as the starting material:[4]

$$C_4H_8S + C_6H_5OH + Cl_2 \xrightarrow{HCl} C_4H_8\overset{\oplus}{S}-C_6H_4-OH + Cl^{\ominus} + HCl \qquad (7\text{-}7)$$

Both methods lead to polymers containing four methylene groups. They cannot be used for the synthesis of polymers with five methylene groups, since the tetrahydrothiopyran ring does not undergo ring-opening under the polymerization conditions (see below, for comparison). The corresponding monomers with five methylene groups are obtained from:[1]

$$Br(CH_2)_5Br + CH_3{-}S{-}C_6H_4{-}OH \longrightarrow (CH_2)_5S^{\oplus}{-}C_6H_4{-}OH + Br^{\ominus} + CH_3Br \qquad (7\text{-}8)$$

The resulting compounds are then converted into zwitterions by treatment with sodium methoxide or anion exchange resins:

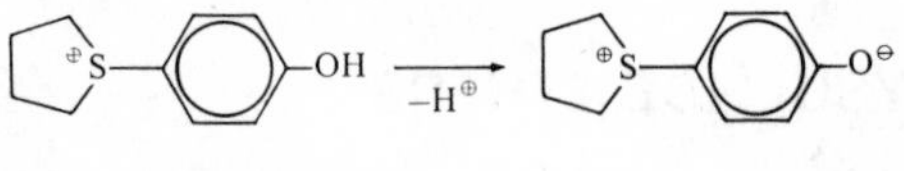

$$(CH_2)_4S^{\oplus}{-}C_6H_4{-}OH \xrightarrow[-H^{\oplus}]{} (CH_2)_4S^{\oplus}{-}C_6H_4{-}O^{\ominus} \qquad (7\text{-}9)$$

These ACSZ monomers (aryl cyclic sulfonium zwitterions) then polymerize with loss of charge and ring-opening to the polysulfide ethers:[1,2]

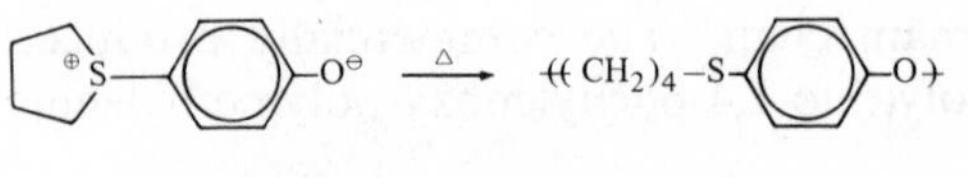

$$(CH_2)_4S^{\oplus}{-}C_6H_4{-}O^{\ominus} \xrightarrow{\Delta} {+\!\!\!(}(CH_2)_4{-}S{-}C_6H_4{-}O{+} \qquad (7\text{-}10)$$

Because of the loss of charge, the reaction is also called a "death-charge" polymerization.

In addition to bifunctional monomers (I–III), also tetrafunctional (IV) and polyfunctional (V) monomers can be made according to this procedure. Chlorophenols do not react with tetrahydrothiophene-1-oxide because of the electron withdrawing effect of the chlorine group. Chlorinated ACSZ monomers (for instance, II) are synthesized, therefore, by the chlorination of the ACSZ monomer itself.

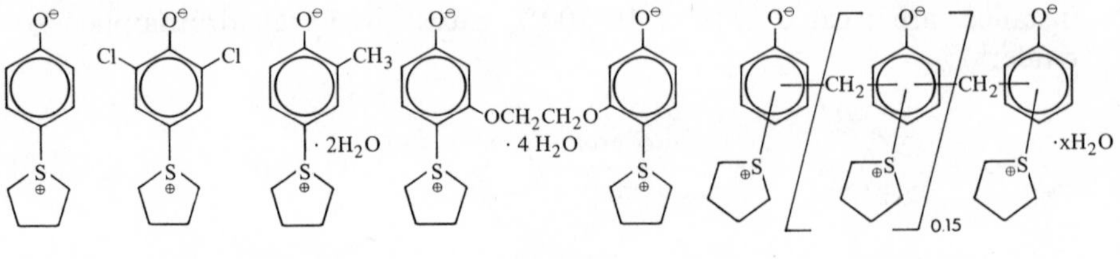

$O^{\ominus}$ Cl Cl CH_3 $\cdot 2H_2O$ OCH_2CH_2O $\cdot 4H_2O$ CH_2 CH_2 $\cdot xH_2O$ S⊕ 0.15

I II III XD-8156 IV XD-8157 V XD-8383

The polymerization of the unchlorinated ACSZ monomers is unusually fast. For instance, the *o*-cresol derivitive (III) polymerizes fully within 30 min at 100°C. Orientation of the polar hydrated zwitterions is probably responsible for this fast polymerization in the crystalline state. The chlorinated species, on the other hand, are not hydrated and are much more stable than the unchlorinated species: they may be stored at room temperature almost indefinitely.

7.3.3 Properties

The viscosities of concentrated solutions of monomers III–V are fairly low: they lie between (4–6) $\times 10^{-3}$ Pa s for a 30% solution and (25–60) $\times 10^{-3}$ Pa s for a 50% solution.[3]

The ACSZ monomers are distinctive in that they can be polymerized from an aqueous solution into a coating which is water resistant. The linear polymers prepared from bifunctional ACSZ monomers *via* phenols (for instance I–III) are relatively low molecular weight and soft. For instance, the polymer from II with a number average molecular weight of 46,000 g/mol has a density of 1.483 g/cm^3, a melting temperature of about 150°C and a glass temperature of about 10°C.[2,4] Published mechanical properties are listed in Table 7-3.

TABLE 7-3
Properties of Poly(thio-(3,5-dichloro)-1,4-phenyleneoxytetramethylene) (Polymer of II), annealed for 48 h at 110°C, then 7 d at 70°C.[2,4]

Property	Physical unit	Value measured: not annealed	annealed
Tensile strength	N/mm^2	20.8	28.9
Impact strength with notch	J/m	872	100
Modulus	N/mm^2	640	900
Ultimate elongation	%	>260	>260
Heat distortion temperature	°C	34	79

Hard coatings are obtained from the polymerization of mixtures of polyfunctional ACSZ monomers (for instance, IV and V). These mixtures polymerize in 5–15 min at 75–200°C from aqueous or basic solutions to give tough, scratch-resistant coatings. Below about 70°C the polymerization is very slow.[3] The coatings adhere well to glass, metals and various plastic films. They are stable to acids and bases.

The hardness and the solvent stability of these cross-linked films increases with increasing cross-link density, that is with increased content of polyfunctional ACSZ monomer in the starting mixture.[3] Films from polymerized monomer V have a Knoop-hardness of 32. These coatings and similar ones from monomer IV do not soften at temperatures below 280°C.

The hardness of coatings can be further increased by cross-linking the ACSZ monomers in the presence of colloids and emulsions.[3] Examples of these are styrene/butadiene latices, terpolymers of acrylic acid, *n*-butyl acrylate and methyl methacrylate and colloidal silicon dioxide (Ludox AS®). For example, coatings prepared from a mixture of III and V (60/40 = *w*/*w*) with 30% Ludox have a Knoop hardness of 80. ACSZ monomers are also supposed to increase the adherance of rubber films when they are added in small amounts to the rubber latices.[10] The toxicity of the ACSZ has only been investigated in part. The wearing of protective glasses is recommended. Prolonged contact of the monomers with the skin causes slight reddening, and minor swelling.[3]

References

1. M. J. Hatch, M. Yoshimine, D. L. Schmidt and H. B. Smith, *J. Amer. Chem. Soc.*, **93**, 4617 (1971).
2. D. L. Schmidt, H. B. Smith, M. Yoshimine and M. J. Hatch, *J. Polymer Sci.* [*Chem.*], **10**, 2951 (1972).
3. D. L. Schmidt, H. B. Smith and W. E. Broxterman, *J. Paint Technol.*, **46**(588), 41 (1974).
4. U.S. Patent 3636052 (January 18, 1972); The Dow Chemical Co.; inv.: M. J. Hatch, M. Yoshimine, H. B. Smith and D. L. Schmidt.
5. U.S. Patent 3660431 (May 2, 1972); The Dow Chemical Co.; inv.: M. J. Hatch, D. L. Schmidt and H. B. Smith.
6. U.S. Patent 3749738 (July 31, 1973); The Dow Chemical Co.; inv.: M. J. Hatch, D. L. Schmidt and H. B. Smith.
7. U.S. Patent 3749739 (July 31, 1973); The Dow Chemical Co.; inv.: M. J. Hatch, D. L. Schmidt and H. B. Smith.
8. U.S. Patent 3767622 (October 23, 1973); The Dow Chemical Co.; inv.: M. J. Hatch, D. L. Schmidt and H. B. Smith.
9. U.S. Patent 3804797 (April 16, 1974); The Dow Chemical Co.; inv.: W. E. Broxterman, D. L. Schmidt and S. Evani.
10. E. P. Plueddeman, lecture at the California Institute of Technology on May 23–25, 1973, cited in Ref. 3.

7.4 POLYETHER SULFONES

7.4.1 Structure

Poly(*p*-phenylene sulfone) or poly(sulfo-1,4-phenylene) (IUPAC) with the base unit $-\!\!(SO_2-C_6H_4)\!\!-$ has a melting temperature of about 520°C,[1,2] consequently it is very difficult to fabricate commercially. Other polysulfones (see Ref. 2) also have high melting temperatures and could hardly be of commercial value. Products which could be fabricated were obtained only after ether groups were introduced to affect flexibility.[3–5]

These polyether sulfones contain sulfone and ether groups in the main chain. Further, all commercial products also have phenylene rings in the main chain; they all are derived from poly(sulfo-1,4-phenylene-oxy-1,4-phenylene) with the base unit

$$-\!\!(SO_2-C_6H_4-O-C_6H_4)\!\!-$$

The following polyether sulfones are currently on the market:

I	$-SO_2-C_6H_4-O-C_6H_4-$	ICI Polyethersulfone 200 P
II	$-SO_2-C_6H_4-O-C_6H_4-$ > $-SO_2-C_6H_4-C_6H_4-$	ICI Polyethersulfone 720 P
III	$-SO_2-C_6H_4-O-C_6H_4-$ < $-SO_2-C_6H_4-C_6H_4-$	3 M Corp. Astrel 360® (now Carborundum)
IV	$-SO_2-C_6H_4-O-C_6H_4-C(CH_3)_2-C_6H_4-O-C_6H_4-$	Union Carbide Corp. Udel®

Udel (IV), also designated as Bakelite Polysulfone,[6,7] was brought on the market in 1965, Astrel 360 in 1967, but the ICI Polyether Sulfones not until 1972.

7.4.2 Synthesis

Three routes are suitable for the synthesis of polyether sulfones. The oxidation of aryl sulfide bridges (see Ref. 2) is already in use for aromatic polysulfides, but apparently not for aromatic polysulfide ethers. The two commercial

$$-\!\!\!+Ar-S+\!\!\!- \longrightarrow -\!\!\!+Ar-SO_2+\!\!\!- \qquad (7\text{-}11)$$

polyether sulfone syntheses are not dependent on this type of polymer-analogous conversion, but rather on polycondensation reactions. In these, either the aryl ether is treated with sulfonyl chloride in the presence of catalysts such as $FeCl_3$, $SbCl_5$, *etc.*, to give sulfone groups[3,5]

$$C_6H_5-O-C_6H_4-SO_2Cl \longrightarrow +C_6H_4-O-C_6H_4-SO_2+ \;+\; HCl \qquad (7\text{-}12)$$

$$C_6H_5-O-C_6H_5 + ClSO_2-C_6H_4-SO_2Cl \longrightarrow +C_6H_4-O-C_6H_4-SO_2-C_6H_4-SO_2+ \;+\; 2\,HCl \qquad (7\text{-}13)$$

or sulfones are treated with phenolates to give ether groups

$$Cl-C_6H_4-SO_2-C_6H_4-OMt \longrightarrow +C_6H_4-SO_2-C_6H_4-O+ \;+\; MtCl \qquad (7\text{-}14)$$

or [4-6]

$$Cl-C_6H_4-SO_2-C_6H_4-Cl + MtO-Ar-OMt \longrightarrow \; -\!\!(C_6H_4-SO_2-C_6H_4-O-Ar-O)\!\!- + 2\ MtCl \quad (7\text{-}15)$$

where Mt is an alkali metal and Ar is an aromatic group, for instance, $-C_6H_4-C(CH_3)_2-C_6H_4-$.

Route (7-12) should be suitable for the synthesis of Polyether Sulfone 200 P (I), since this reaction gives almost exclusively *para*-substitution, whereas route (7-13) gives about 80% *para*- and 20% *ortho*-substitution.[8] Branching occurs more readily in the *ortho*-position. *Ortho*-substitution is undesirable since increasing *ortho*-content leads to more brittle products. But since the starting monomer 4-chlorosulfonyldiphenyl ether for route V (7–12) is too expensive, route (7–14) is used for the synthesis of I. Route (7–14) and the commercially equally feasible route (7–15) yield polymers with reactive end-groups. During fabrication in the melt these end-groups can subsequently condense, leading to an increase in the molecular weight and melt viscosity. Consequently, in order to stabilize the melt viscosity, the phenolic end-groups are treated with methyl chloride to give methoxy groups which are not reactive under these conditions (see also Refs. 8 and 16).

Copolymers II and III are available *via* a number of routes. Ether formation (reaction (7–14) and (7–15)) is preferred for product II, and sulfone formation (reactions (7–12) and (7–13)) is preferred for product III. In both cases these are specialty intermediates, consequently, as a matter of course, copolymers II and III are more expensive than the homopolymers I and IV. For instance, in 1974 II was available from ICI at a price of 55$/kg, whereas I cost 11$/kg.[9]

The Udel-"Polysulfone" (IV) is produced by the reaction (7–15).[6,7]

7.4.3 Properties

All the polyether sulfones are high temperature engineering plastics. They are stable to oxygen and to thermal degradation, show good electrical insulation properties, have high mechanical values and show desirable flame-resistant behavior. Characteristic properties of polyether sulfones I, III and IV are listed in Table 7-4. More detailed data, especially on the long term stability, are found in company literature[9] and in some publications, for instance for Astrel 360 (III) in Ref. 10, for Udel (IV) in Refs. 7 and 11, and for the ICI Polyether Sulfones I and II in Refs. 12 and 13.

TABLE 7-4
Properties of polyether sulfones.

Property	Physical unit	Values measured for: ICI PES 200 P Ref. 9	3 M Astrel 360 Ref. 10	Union Carbide Udel Refs. 6–8, 10, 18
Density	g/cm^3	1.37	1.36	1.24
Glass temperature	°C	230	288	190–195
Heat distortion temperature (at 1.82 N/mm^2)	°C	203	274	174
Coeff. of linear thermal expansion	K^{-1}	5.5×10^{-5}	4.7×10^{-5}	5.6×10^{-5}
Specific heat capacity	$J\,g^{-1}K^{-1}$	1.12		
Thermal conductivity	$W\,m^{-1}K^{-1}$	0.18		
Tensile yield strength	N/mm^2	84	90	72
Tensile strength	N/mm^2		162	
Modulus	N/mm^2	2440	2600	2520
Yield	%		13	
Compressive strength	N/mm^2		126	
Compressive modulus	N/mm^2		2400	
Flexural strength	N/mm^2	130	121	110
Flexural modulus	N/mm^2	2570	2780	2750
Impact strength with notch	J/m	87	164	69
Rockwell hardness	—	M88	M110	M69
Dielectric constant	—	3.5	3.5	3.0
Volume resistivity	Ω cm	10^{17}–10^{18}	3×10^{16}	5×10^{16}
Power factor (60 Hz)	—	0.001		
(10^6 Hz)	—	0.006		
Dissipation factor	—		0.0030	0.0008–0.0034
Arc resistance	s		67	122
Flammability	—	not flammable	self-extinguishing	
Water absorption (24 h)	%	0.43	1.8	0.3
Equilibrium water content, 20 °C	%	2.1		
100 °C	%	2.3		

All the polyether sulfones have excellent creep resistance (Figure 7-5). They are unaffected by hydrocarbons and aqueous acids and bases, and only slightly affected by alcohols and detergent solutions. A few ketones induce stress cracking. Dimethyl formamide, dimethyl sulfoxide and *N*-methylpyrrolidone are solvents. Conventional injection molding machines can be used to fabricate the polyether sulfone I and IV, but polyether sulfone II only with difficulty. However, the high glass temperature and melt viscosity of polyether sulfone III (see Table 7-4) make special equipment necessary for

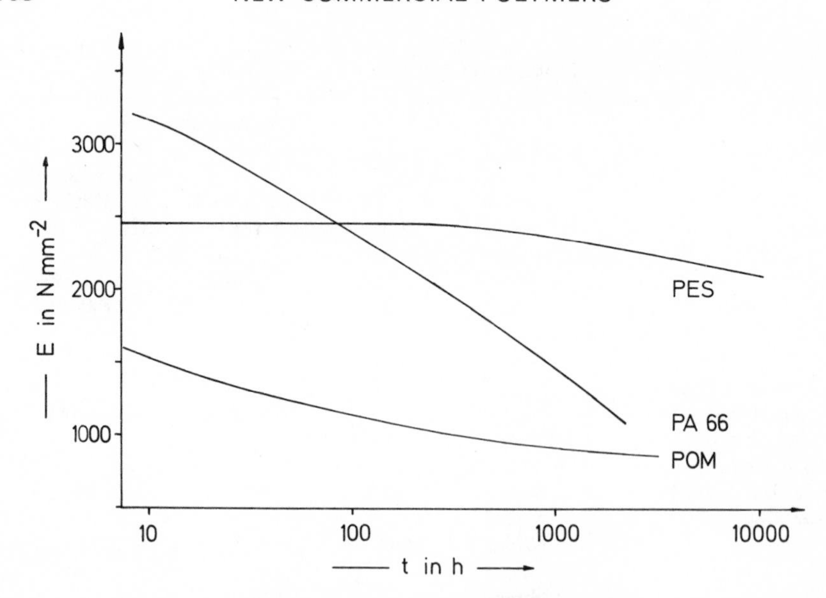

Figure 7-5 Creep modul *E* of Polyether Sulfone 200 P, a polyamide 6,6 and a poly(oxymethylene) at 20 °C as a function of time *t* (Ref. 9).

its fabrication. These special injection molding machines must permit material temperatures of about 425°C and injection pressures of 1400–2800 kg/cm^2. The equipment temperature should be about 205–260°C. In principle, all the polyether sulfones can be extruded, at least the usual types.

7.4.4 Application

Polyether sulfones are suitable for engineering and electrical parts, for example as electrical terminal holders, switches, valve bodies, bearings, automobile parts, printed circuits, fuel-cell housings, household utensils *etc*. Descriptions of special applications are found in the literature (see Refs. 9–12, 14, 15 and 17).

References

1. British Patent 1234008; German Patent 1938806 (February 12, 1970); Inventa AG; inv.: J. Studinka and R. Gabler, *C. A.*, **72**, 91416R (1970).
2. R. Gabler and J. Studinka, *Chimia* [*Aarau*], **28**, 567 (1974).
3. British Patent 1060546 (U.S. Pat. Appl. April 16, 1963); 3M Corp.; inv.: H. A. Vogel; *C. A.*, **67**, 22350z (1967).
4. British Patent 1078234; Dutch Pat. Appl. 6408130 (U.S. Pat. Appl. July 16, 1963); Union Carbide Corp.; inv.: A. G. Farnham and R. N. Johnson; *C. A.*, **63**, 1949b (1965).

5. British Patent 1016245, Belg. Pat 639634 (Brit. Appl. November 6, 1962); inv.: M. E. B. Jones; *C. A.*, **63**, 700f (1965).
6. R. N. Johnson, A. G. Farnham, R. A. Clendinning, W. F. Hale and C. N. Merriam, *J. Polymer Sci.* [*A*-1], **5**, 2375 (1967).
7. W. F. Hale, A. G. Farnham, R. N. Johnson and R. A. Clendinning, *J. Polymer Sci.* [*A*-1], **5**, 2399 (1967).
8. J. B. Rose, *Chimia* [*Aarau*], **28**, 561 (1974).
9. Product information literature of the firm ICI, England and ICI, United States; P-1 Rev. 11/1/74, PES 10(33/972), PES 101 and PES 102.
10. G. Morneau, *Mod. Plastics*, **48**(1), 150 (1970).
11. T. E. Bugel and R. K. Walton, *Machine Design*, **37**, 195 (1965).
12. V. J. Leslie, J. B. Rose, G. O. Rudkin and J. Feltzin, *ACS Div. Org. Coatings Plast. Chem. Pap.*, **34**(1), 142 (1974); *ibid.*, in R. D. Deanin, ed., *New Industrial Polymers, ACS Symp. Ser.*, **4**, 63 (1974).
13. K. V. Gotham and S. Turner, *Polymer* [*London*], **15**, 665 (1974).
14. ——, *Kunststoffe*, **60**, 478 (1970).
15. ——, *Mod. Plastics*, **51**(3), 51 (1974).
16. J. B. Rose, *Polymer* [*London*], **15**, 456 (1974).
17. W. R. Bergenn, *Plastics Engng.*, **31**, 28 (1975).
18. Product information literature of Union Carbide.

8 Aliphatic Polyamides

8.1 POLY(ALPHA-AMINO ACIDS)

8.1.1 Structure

Poly(α-amino acids) with the base unit

$$\text{+NH—CO—CHR+}$$

are formally derivatives of polyamide 2, that is, of poly(glycine) ($R=H$). When $R \neq H$, the α-amino acids may occur in either a D- or L-configuration. As a result, it is possible to obtain configurational copolymers in addition to the two homopolymers. However, such copolymers made from D- and L-units do not have the good mechanical properties of the homopolymeric poly(L-amino acids) and poly(D-amino acids), and, therefore, they have attained no commercial significance.

The firm Courtaulds commercially produced poly(L-alanine) (where $R=CH_3$) and poly(γ-methyl-L-glutamate) (where $R=CH_2CH_2COOCH_3$) in experimental quantities 20 years ago.[1-4] Production was terminated because of the high price of the corresponding pure antipode-monomer at that time. However, the price is dropping with increasing α-amino acid production (Figure 8-1). Currently L-glutamic acid is marketed for about 1\$/kg,[5] poly(γ-methyl-L-glutamate) films for about 4.30\$/kg in small quantities f.o.b. Hamburg.[7] Poly(γ-methyl-L-glutamate) is used in Japan as synthetic leather.[4] Poly(L-glutamic acid) is a promising candidate for silk-like fibers.[8]

8.1.2 Synthesis

Currently α-amino acids are produced on a commercial scale by a number of routes:[5,9] by the fermentation of glucose (see also Refs. 10 and 11), enzymatically from various substances, by the hydrolysis of proteins, synthetically *via* the Strecker synthesis or by conversion from other α-amino acids (Table 8-1). All of the procedures, with the exception of the wholly synthetic ones, give the L-amino acids. The racemate which results from the fully synthetic method is separated by physical means and the D-amino acid reracemized. The asymmetric hydrogenation of double bonds with, for instance, chiral

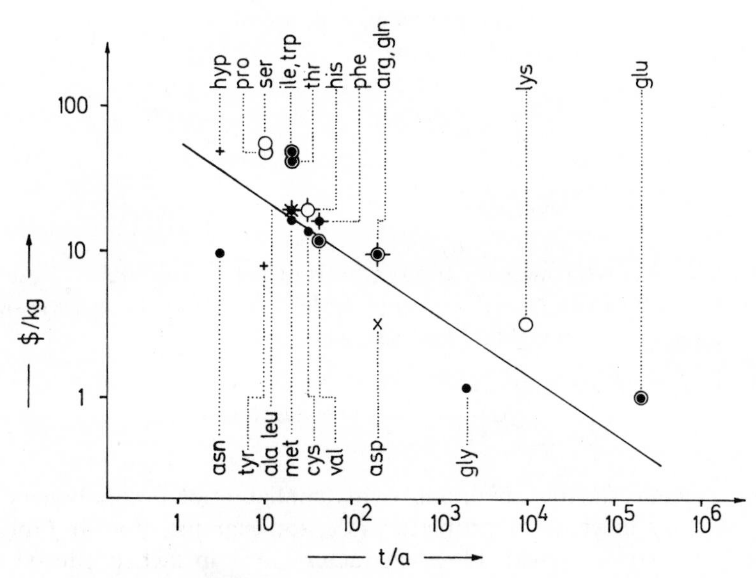

Figure 8-1 Experience diagram for Japanese α-amino acid production in 1973 (according to figures from Ref. 5, see also Ref. 6). The prices in yen were converted to U.S.-dollars based on conversion rates in effect in late 1974. ● wholly synthetic, ○ fermentation, × enzymatic, + hydrolysis and extraction of proteins. (For an explanation of symbols, see Table 8-2.)

TABLE 8-1

Commercial-scale syntheses of α-amino acids (compare with Refs. 5, 6, 11 and 15); for the definition of symbols, see Table 8-2.

Method	α-amino acids produced	Configuration
Fermentation of glucose	arg, asp, gln, glu his, ile, lys, pro, val, thr	L
of glycine	ser	L
Enzymatically from asn	ala	L
from fumaric acid	asp	L
Hydrolysis of proteins	arg, cys/cys, his, hyp, leu, tyr	L
Wholly synthetic from aldehydes	ala, gly, leu, met, phe, thr, trp, val	D, L
from acrylonitrile	glu, lys	D, L
from caprolactam	lys	D, L
Conversion from other α-amino acids		
from orn	arg	L
from cys/cys	cys	L
from glu	gln	L

Rh(I)-complexes,[12–14] which proceeds with about 90% optical yield is not yet being used on a commercial scale.

$$\mathrm{R{-}CH{=}C(COOR')(NHCOCH_3)} + \mathrm{H_2} \longrightarrow \mathrm{R{-}CH_2{-}C(COOR')(NHCOCH_3){-}H} \qquad (8\text{-}1)$$

Of the many known routes for the synthesis of poly(α-amino acids) only the polymerization of the *N*-carboxy anhydrides[16] gives products with sufficiently high molecular weights[17] (see also Refs. 1–4).

$$\mathrm{RCH{-}CO{-}O{-}CO{-}NH} \text{ (ring)} \longrightarrow \mathrm{{+}NH{-}CO{-}CHR{+}} + \mathrm{CO_2} \qquad (8\text{-}2)$$

Use of racemic starting materials yields configurational copolymers with unsatisfactory mechanical properties (see, for example, Ref. 4). Optically active polymers[18–25] with block-character[25] are, in fact, produced from racemic *N*-carboxy anhydrides by using optically active initiators. However, this method of racemate separation *via* polymerization is as yet of no commercial importance.

8.1.3 Properties and Fabrication

Poly(α-amino acid) molecules can appear in two conformational states in the crystal. The α-form is a helix with about 3.7 amino acid residues per turn which is stabilized by intramolecular hydrogen-bonds and by hydrophobic bonding. The β-form is the so-called pleated-sheet structure; it is insoluble and unmeltable as a result of the numerous intermolecular hydrogen-bonds. The α-form gives wool-like fibers, the β-form gives silk-like products.[4]

The preferred conformations of the various poly(α-amino acids) are listed in Table 8-2. The β-form is soluble only in those solvents which can break the existing hydrogen-bonds, and it then assumes the form of a random coil. The α-form retains its rod-like shape in the so-called helicogenic solvents, whereas in other solvents it also assumes the shape of random coils.

Poly(α-amino acid) film- and fiber-formation are influenced by the polymer's molecular weight, conformation and association of the polymers. Poly(α-amino acids) are not processable in the melt, since they degrade just below the melting point. Fiber-formation by solution-spinning requires quite high molecular weights. On the other hand, films can be made from low molecular poly(α-amino acids). As a rule, polymerization in the β-form gives only low molecular weights.[4,26] Consequently, α-amino acids such as

TABLE 8-2

Names, structures and symbols for naturally occurring α-amino acids, $NH_2-CHR-COOH$, as well as the conformations of the corresponding poly(α-amino acids) in the solid state, α=α-helix, β=pleated-sheet structure, [ ]=conformation after stretching.

*Iminocarboxylic acid with the structure complete as drawn.

Symbol	Name	R	Conformation of poly(α-amino acid)
ala	Alanine	CH_3	α[β]
arg	Arginine	$(CH_2)_2N{=}C(NH_2)_2$	—
asn	Asparagine	CH_2CONH_2	—
asp	Aspartic acid	CH_2COOH	α
cys	Cysteine	CH_2SH	β
cys/cys	Cystine	$CH_2-S-S-CH_2$	—
gln	Glutamine	$(CH_2)_2CONH_2$	—
glu	Glutamic acid	$(CH_2)_2COOH$	α
gly	Glycine	H	β
his	Histidine	CH_2- N N	α
hyp	Hydroxyproline*	$H_2\overset{\oplus}{N}$ $COO^{\ominus}$ OH	3_1-Helix
ile	Isoleucine	$CH(CH_3)CH_2CH_3$	β
leu	Leucine	$CH_2CH(CH_3)_2$	α
lys	Lysine	$(CH_2)_4NH_2$	α[β]
met	Methionine	$(CH_2)_2SCH_3$	α
orn	Ornithine	$(CH_2)_3NH_2$	—
phe	Phenylalanine	$CH_2C_6H_5$	α
pro	Proline*	$H_2\overset{\oplus}{N}$ $COO^{\ominus}$	10_3-or 3_1-Helix
ser	Serine	CH_2OH	β
thr	Theonine	$CH(CH_3)OH$	β
trp	Tryptophane	CH_2 N	α
tyr	Tyrosine	$CH_2-C_6H_4OH$	α
val	Valine	$CH(CH_3)_2$	β

cysteine, glycine, isoleucine, serine, threonine and valine (see Table 8-2) do not produce fiber-forming polymers. Spinning of poly(α-amino acids) in the α-form also presents problems, since many solutions form mesomorphous phases, even at concentrations of 10–15%. These are highly viscous and are not spinnable with normal spinning techniques. The structure in these solutions can be disrupted by the addition of helix-breaking solvents, such as dichloroacetic acid. These poly(α-amino acid) solutions are then spin-

TABLE 8-3

Mechanical properties of various poly(α-amino acids) according to data from Ref. 4 and Ref. 8. (For an explanation of the symbols, see Table 8-2.)

Property	Physical unit	Poly(α-amino acid)						
		natural silk	*p*(glu)	*p*(ala)	*p*(*m*-glu)	*P*(leu) silk	*p*(leu) wool	natural wool
Density	g cm^{-3}	1.37	1.46	1.26	1.31	1.026	1.037	1.32
Stretching ratio			2–3	2.0	1.7	1.9		
Titer (tex)	g km^{-1}			0.22	0.23	0.20	0.29	0.73
Tensile strength								
dry	g tex^{-1}	36	21	39	20	19	5	14
wet	g tex^{-1}	37	16	25	18	18	6	
Elongation								
dry	%	20	20[a]	9.3	19	17	55	41.2
wet	%	22	13[a]	11.9	17	20	97	
Modulus								
dry	g tex^{-1}	900	280[b]	920	370	315	151	229
wet	g tex^{-1}	765	153[b]					
Recovery after 3% elongation	%	64		100	94		16.8	93.5
Water absorption at 80% relative humidity at 20 °C	%	13.5		8.7	2.4			

[a]Stretching ratio 1:1.65.
[b]Stretching ratio 1:1.54.

nable, because of the polymer's coil structure, providing the molecular weight of the polymer is high enough. Corrosion problems with the spinning apparatus appear when dichloroacetic acid is used as the solvent.

One poly(α-amino acid) conformation can be transformed into the other by treatment with suitable solvents by stretching or by thermal treatment in the presence or absence of solvent (see Table 8-2). Stretching produces largely silk-like fibers, whereas thermal treatment (shrinking) gives predominantly wool-like fibers.

Poly(L-alanine) in the β-conformation yields a fiber that most closely resembles silk. Poly(L-leucine) contains both the α- as well as the β-conformation after stretching. Poly(L-leucine) looks like silk, but has mechanical properties which lie between that of silk and wool. Poly(γ-methyl-L-glutamate) with largely a β-conformation is also silk-like (Table 8-3).

Poly(L-alanine) and poly(L-glutamic acid) yellow less than raw silk. Poly(L-alanine) burns without first melting. The decomposition temperature of poly(L-glutamic acid) is about 270°C. Poly(L-glutamic acid) is slowly decomposed in the ground by microorganisms.

Poly(L-glutamic acid), in accordance with its chemical structure, can be dyed with cationic dyes to give brilliant shades. Poly(γ-methyl-L-glutamate) can be dyed with both cationic and dispersion dyes; however, it is difficult to obtain deep shades.

References

1. C. A. Bamford, A. Elliot and W. E. Hanby, *Synthetic Polypeptides*, Academic Press, New York, 1956.
2. D. G. H. Ballard, "Synthetic polypeptides", in H. F. Mark, S. M. Atlas and E. Cernia (Eds.): *Man-Made Fibers*, Vol. 2, Interscience, New York, 1968.
3. J. Noguchi, *Chem. High Polymers* [*Tokyo*], **27**, 145 (1970).
4. J. Noguchi, S. Tokura and N. Nishi, *Angew. Makromol. Chem.*, **22**, 107 (1972).
5. S. Mori, *Kobunshi* (*High Polymers Japan*), **22**(250), 31 (1973).
6. H.-G. Elias, *Chem. Techn.*, **5**, 748 (1975); **6**, 244 (1976).
7. G. Ebert, C. Ebert, V. Kroker and W. Werner, *Angew. Makromol. Chem.*, **40/41**, 493 (1974).
8. S. Oya and J. Takahashi, *Chem. Tech*, **3**, 672 (1973).
9. T. Kaneka, Y. Izumi, I. Chibata, T. Itoh, *Synthetic Production and Utilization of Amino Acids*, Kodansha, Tokyo, and Halsted, New York, 1974.
10. D. Perlman, *Chem. Tech.*, **4**, 210 (1974).
11. K. Yamada, S. Kinoshita, T. Tsunoda, K. Aida (Eds.), *The Microbial Production of Amino Acids*, Kodansha, Tokyo, and Halsted, New York, 1974.
12. W. S. Knowles, M. J. Sabacky, B. D. Vineyard, *Chem. Tech.*, **2**, 590 (1972).
13. H. B. Kagan and T. P. Deng, *J. Amer. Chem. Soc.*, **94**, 6429 (1972).
14. W. S. Knowles, M. J. Sabacky and B. D. Vineyard, *Adv. Chem. Ser.*, **132**, 274 (1974).
15. H.-G. Elias, *Makromoleküle*, Hüthig and Wepf, Basel, 3rd ed., 1975.
16. H. Leuchs, *Chem. Ber.*, **39**, 857 (1906).
17. R. B. Woodward and C. H. Schramm, *J. Amer. Chem. Soc.*, **69**, 1552 (1947).

18. T. Tsuruta, S. Inoue and K. Matsuura, *Makromol. Chem.*, **63**, 219 (1963).
19. T. Tsuruta, K. Matsuura and S. Inoue, *Makromol. Chem.*, **80**, 149 (1964).
20. T. Makino, S. Inoue and T. Tsuruta, *Makromol. Chem.*, **150**, 137 (1971).
21. H. G. Bührer and H.-G. Elias, *Makromol. Chem.*, **169**, 145 (1973).
22. S. Yamashita and H. Tani, *Macromolecules*, **7**, 406 (1974).
23. S. Yamashita, K. Waki, N. Yamawaki and H. Tani, *Macromolecules*, **7**, 410 (1974).
24. S. Yamashita, N. Yamawaki and H. Tani, *Macromolecules*, **7**, 724 (1974).
25. H.-G. Elias, H. G. Bührer and J. Semen, *Appl. Polymer Symp.*, **26**, 10 (1975).
26. J. Noguchi, *Progr. Polymer Sci. Japan*, **5**, 65 (1973).

8.2 POLYAMIDE 4

8.2.1 Structure

Polyamide 4, also called nylon 4 or poly(pyrrolidone), has the base unit

$$\leftarrow NH-CO-CH_2CH_2CH_2 \rightarrow$$

and according to IUPAC nomenclature is named poly(iminocarbonyltrimethylene). The history of polyamide 4 is uneven.[1] Initial observations made on vinyl pyrrolidone synthesis at General Aniline and Film led the investigators to the discovery of the polymerization of pyrrolidone[2] at their subsequent place of employment, Arnold, Hoffman and Co. The polymerization of pyrrolidone is accelerated by acyl compounds.[3] The polyamide 4's prepared from this or a variety of other co-activators or stabilizers (see Ref. 1) decompose with depolymerization at the melting point. Consequently, they cannot be melt-spun. However, anionic polymerization of pyrrolidone with, for instance, sodium pyrrolidone as initiator and carbon dioxide as co-initiator[4,5] supposedly gives a satisfactory, thermally stable product (compare Refs. 6 and 7). Its production under the tradename Tajmir® was announced by the firm Alrac after extensive testing at the Southern Research Institute. The nylon 4 patents were assumed by Vestra Corp. (Chevron Chemical Co.) in January 1975.[8]

8.2.2 Synthesis

Thermodynamic considerations (compare, for example, Ref. 9) and experimental observations on the degradation of nylon 4 show that the ceiling-temperature for the monomer/polymer polymerization equilibrium must be fairly low. Although this is the decisive quantity for the synthesis of polyamide 4, there is, nevertheless, still no published data available on it.

Pyrrolidone polymerization is initiated by sodium or potassium pyrrolidone with *N*-acetylpyrrolidone or its precursors as activators. It probably proceeds analogously to that known for other cyclic amides. The initiation step

$$\text{X–CO–N}\langle\text{pyrrolidone}\rangle + {}^{\ominus}\text{N}\langle\text{pyrrolidone}\rangle \longrightarrow \text{X–CO–}\overset{\ominus}{\text{N}}\text{–(CH}_2)_3\text{–CO–N}\langle\text{pyrrolidone}\rangle \quad (8\text{-}3)$$

is followed by regeneration of the initiator

$$\text{X–CO–}\overset{\ominus}{\text{N}}\text{–(CH}_2)_3\text{–CO–N}\langle\text{pyrrolidone}\rangle + \text{HN}\langle\text{pyrrolidone}\rangle \longrightarrow \text{X–CO–NH–(CH}_2)_3\text{–CO–N}\langle\text{pyrrolidone}\rangle + {}^{\ominus}\text{N}\langle\text{pyrrolidone}\rangle \quad (8\text{-}4)$$

and finally the actual propagation step

$$\text{X–CO–NH–(CH}_2)_3\text{–CO–N}\langle\text{pyrrolidone}\rangle + {}^{\ominus}\text{N}\langle\text{pyrrolidone}\rangle \longrightarrow \text{X–CO–NH–(CH}_2)_3\text{–CO–}\overset{\ominus}{\text{N}}\text{–(CH}_2)_3\text{–CO–N}\langle\text{pyrrolidone}\rangle \quad (8\text{-}5)$$

The reaction sequence of equations (8-4)/(8-5) is then repeated. It is speculated that when *N*-acetylpyrrolidone ($X{=}CH_3$) is the co-initiator a

$$\text{X–CO–NH}\sim\sim + {}^{\ominus}\text{N}\langle\text{pyrrolidone}\rangle \longrightarrow \text{X–CO–N}\langle\text{pyrrolidone}\rangle + {}^{\ominus}\text{NH}\sim\sim \quad (8\text{-}6)$$

trans-initiation also occurs.[1] This *trans*-initiation is thought to regenerate co-initiator which in turn would continually form new polymer molecules, even at extremely low initiator concentrations. Also, it could cause the experimentally observed broad molecular weight-distribution. In the "new" polyamide 4-polymers, however, $X{=}O^{\ominus}$, that is, the polymerization is initiated with alkali metal pyrrolidone/CO_2, and *trans*-initiation would practically be excluded because of repulsion of negative charges in the transition state. In fact, these "new" polyamide 4-polymers have a narrower molecular weight-distribution than the "old" polymers initiated with alkali metal pyrrolidone/*N*-acyl compounds. Instead of CO_2, SO_2 can also be used.[10]

8.2.3 Properties

The "old" polyamide 4-polymers decompose thermally by the successive splitting out of pyrrolidone units, that is by depolymerization.[1] At the same time the polymer browns and smoke is evolved. The "new" polyamide 4-polymers, on the other hand, decompose randomly with chain-splitting. In contrast to the "old" PA 4-polymers, they can be melt-spun.[1]

The melting point of polyamide 4-polymers is 260–265°C.[11,12] The degree of crystallinity is not known. The mechanical properties correspond to those of polyamide 6 (Table 8-4). However, the textile properties are similar to those of cotton; that is, the fibers from PA 4 have a quite high water absorption, good dyeability and good light stability. Textiles from PA 4 undergo at least 50 washings without loss of the properties of the material; the "new" PA 4 does not fibrilate in contrast with the "old" PA 4.

TABLE 8-4
Properties of polyamide 4.

Property	Physical unit	Polyamide 4 "old" Refs. 10 and 12	Polyamide 4 'new" Ref. 1
Density	g/cm^3		1.25
Stretching ratio		3	3.5
Tenacity	km		40.5
Modulus	g/tex		171
Ultimate elongation	%	42–71	36
Knot strength	g/tex	16	37
Water absorption (20 °C, 65% relative humidity)	%	8.4 9.1	10

PA 4 occupies, therefore, a mid-position between the more hydrophobic synthetic fibers and the more hydrophilic natural fibers, a fact which explains the textile industry's interest in these polymers. PA 4 has also been suggested as a raw material for synthetic leather.[6,7]

References

1. E. M. Peters and J. A. Gervasi, *Chem. Tech.*, **2**, 16 (1972).
2. U.S. Patent 2638463 (December 7, 1951); Arnold, Hoffman and Co., inv.: W. D. Ney, Jr., W. R. Nummy and C. E. Barnes; *C. A.*, **47**, 9624c (1953).
3. U.S. Patent 2739959 (February 24, 1953); Arnold, Hoffman and Co.; inv.: W. D. Ney, Jr. and M. Crowther; *C. A.*, **50**, 13504d (1956).
4. German Pat. Appl. 1911834 (March 8, 1969); Radiation Research Corp.; inv.: C. E. Barnes; *C. A.*, **74**, 4556m (1971).
5. Anonym., *Chem. Week*, **16**, 67 (1970).
6. Anonym., *Mod. Text.*, **50**(12), 39 (1969).
7. Anonym., *Chemiefasern*, **20**(1), 30 (1970).
8. Private communication, R. E. Fiertz, Vestra Corp. (March 10, 1975).
9. H.-G. Elias, *Makromoleküle*, Chapter 16, Hüthig and Wepf, Basel, 3rd ed., 1975.
10. G. Schirawski, *Makromol. Chem.*, **161**, 57 (1972).
11. E. Müller, *Melliand Textilber.*, **44**, 484 (1963).
12. K. Dachs and E. Schwartz, *Angew. Chem.*, **74**, 540 (1962); *Angew. Chem. Internat. Ed. Engl.*, **1**, 430 (1962).

8.3 POLYAMIDE 6,12

Polyamide 6.12 has the repeat unit

$$-NH-(CH_2)_6-NH-OC-(CH_2)_{10}-CO-$$

It is produced from hexamethylenediamine and 1,10-decanedicarboxylic acid by DuPont. The same C_{12}-acid is used in the production of Qiana® (see Chapter 8.5). DuPont expects that this new product will gradually replace Nylon 6.10.

Polyamide 6.12 has good dimensional stability. It absorbs less water than Nylon 6.10 and is also stiffer (see Table 8-5); however, it can be processed in the same way. DuPont is also offering glass fiber reinforced varieties. The material is self-extinguishing (ASTM D-635).

TABLE 8-5
Properties of polyamide 6.12.[1]

Property	Physical unit	Values measured for Common	Glass fiber reinforced 33%	43%
Density	g/cm³	1.06–1.08	1.32	1.48
Melting temperature	°C	218		
Heat distortion temperature at 1.86 N/mm²	°C	66	210	210
Coefficient of linear thermal expansion	K^{-1}	9×10^{-5}	2.3×10^{-5}	2.2×10^{-5}
Tensile strength	N/mm²	62	169	197
Ultimate elongation	%	100	4.5	4.0
Flexural modulus	N/mm²	2500	8500	10,500
Shear strength	N/mm²	59	77	84
Impact strength	J/m	46	142	180
Deformation under load				
(14 N/mm² at 50 °C)	%	1.6		
(28 N/mm² at 50 °C)	%		1.0	0.5
Rockwell hardness	—	R-114	R-118	R-118
Dielectric constant	—	3.6	3.7	4.0
Volume resistivity	Ω cm	10^{15}	1.4×10^{15}	1.1×10^{15}
Water absorption (24 h)	%	0.4	0.20	0.15

References

1. Anonym., *Modern Plastics*, **47**(10), 107 (1970).

8.4 POLY(TRIMETHYLHEXAMETHYLENETEREPHTHALAMIDE)

8.4.1 Structure and Synthesis

This polymer is made up of terephthaloyl units (I) and trimethylhexamethylenediamine units (II, III)

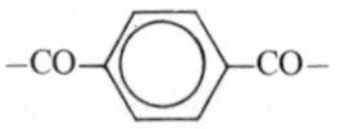

$$-CO-C_6H_4-CO- \qquad \text{I}$$

$$-NH-CH_2-C(CH_3)_2-CH_2-CH(CH_3)-CH_2-CH_2-NH- \qquad \text{II}$$

$$-NH-CH_2-CH(CH_3)-CH_2-C(CH_3)_2-CH_2-CH_2-NH- \qquad \text{III}$$

The product was developed in the Zurich research laboratories of W. R. Grace and Co.[1] and later produced under license by Dynamit Nobel AG.[2,3] The commercial product, which presumably is made up in the ratio 2:1:1 (I:II:III), is sold under the tradename Trogamid T®.

The diamines II and III are obtained from acetone *via* isophorone and the corresponding dicarboxylic acids and dinitriles:

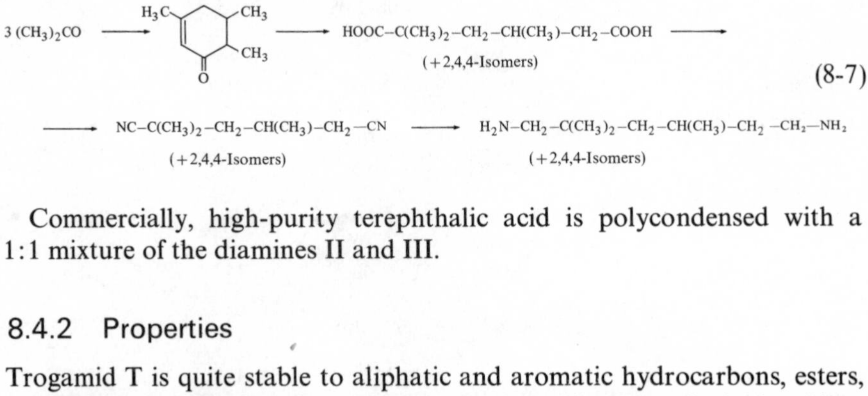

$$3\,(CH_3)_2CO \longrightarrow \text{isophorone} \longrightarrow \underset{(+2,4,4\text{-Isomers})}{HOOC-C(CH_3)_2-CH_2-CH(CH_3)-CH_2-COOH} \longrightarrow \qquad (8\text{-}7)$$

$$\longrightarrow \underset{(+2,4,4\text{-Isomers})}{NC-C(CH_3)_2-CH_2-CH(CH_3)-CH_2-CN} \longrightarrow \underset{(+2,4,4\text{-Isomers})}{H_2N-CH_2-C(CH_3)_2-CH_2-CH(CH_3)-CH_2-CH_2-NH_2}$$

Commercially, high-purity terephthalic acid is polycondensed with a 1:1 mixture of the diamines II and III.

8.4.2 Properties

Trogamid T is quite stable to aliphatic and aromatic hydrocarbons, esters, salt solutions, dilute alkali and dilute mineral acids. It has limited stability to aldehydes, ketones and lower aliphatic alcohols. Aniline, pyridine and tetrahydrofuran decompose it. Phenols, formic acid, conc. sulfuric acid and dimethylformamide are solvents. Trogamid is not unaffected by hot water.[4] It absorbs less water than polyamide 6 and polyamide 66 and only slightly more water than polyamide 11;[3] water, however, is not a plasticizer.[4]

The homo- and copolymers of terephthalic acid with the trimethylene-

hexamethylenediamines have only a low crystalline order.[1] Spherulites and other supermolecular structures are not found. The polymer is considered X-ray amorphous. Still, surprisingly for these types of polymers, "melting points" are reported; a contradiction, since only crystalline materials can have melting points. These "melting points" are 220–228 °C.[1] It is unclear just what physical transformation is occurring at this temperature. Differential thermal analysis measurements give values of 149[1] or 145–153 °C[3] as the glass temperatures, which are also supported by the data from mechanical damping.[2] The stress/strain curve shows after a steep climb a pronounced

TABLE 8-6

Properties of dry Trogamid® at 20 °C.[2,4,5]

Property	Physical unit	Values measured for T (unfilled)	TG 35 (35% glass fibers)
Density	g/cm³	1.12	1.38
Refractive index	F	1.566	
Brittleness temperature	°C	−12	
Heat distortion temperature: Martens	°C	100	130
Vicat	°C	145	145
ISO/R 75 (1.86 N/mm²)	°C	130	135
Coefficient of linear thermal expansion	K^{-1}	6×10^{-5}	7×10^{-5}
Specific heat capacity	$J\,g^{-1}\,K^{-1}$	1.46	1.25
Thermal conductivity	$kJ\,m^{-1}\,h^{-1}\,K^{-1}$	0.75	0.84
Tensile yield strength	N/mm²	85	
Yield	%	9.5	
Modulus	N/mm²	3000	10,000
Tensile strength	N/mm²	60	140
Ultimate elongation	%	70	3
Compressive strength	N/mm²	120	
Compressive deformation	%	11.5	
Flexural strength	N/mm²		170
Impact strength 20 °C	kJ/m²	no break	30
−50 °C	kJ/m²	60	
Impact strength with notch 20 °C	kJ/m²	10–15	7
−50 °C	kJ/m²	3–5	
Hardness, indentation	N/mm²	125	170
Rockwell	—	M 93	
Abrasion resistance	mg/100 U	21–25	
Dielectric constant	—	3.5	4.1
Volume resistivity	Ω cm	$> 10^{14}$	10^{16}
Surface resistivity	Ω	$> 10^{13}$	10^{13}
Dissipation factor	—	0.03	0.021
Dielectric strength	kV/mm	25	30

yield point at only small elongation. The high yield point and the high modulus serve to identify Trogamid T® as an engineering material (see Table 8-6). The modulus varies only slightly between −60°C and +140°C, to just under the glass temperature. After 90 days storage at 90°C unstabilized polymers show slight embrittling, that is the impact strength with notch decreases, and the tensile stress at maximum load, the flexural stress at yield and hardness increase. Addition of stabilizers raises thermal stability to +130°C.[2] Trogamid T® also has good long-term behavior as is shown by its deformation behavior under time-constant loading (Figures 8-2 and 8-3).

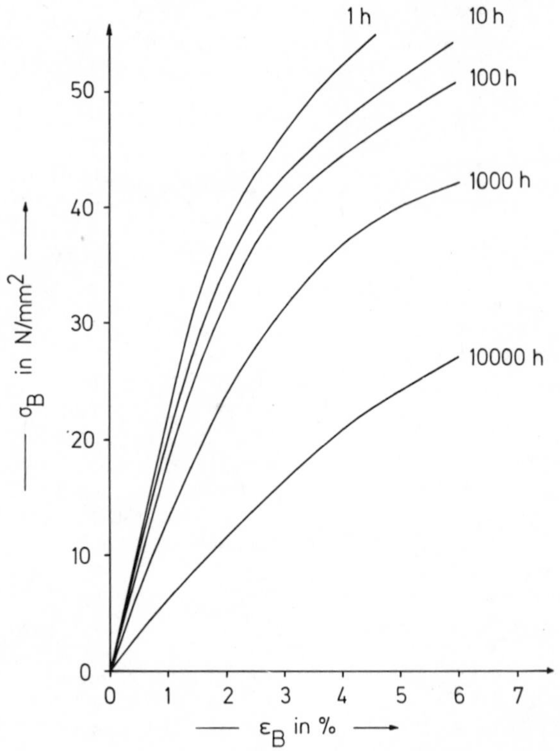

Figure 8-2 Isochronal curves of tensile strength *vs*. ultimate elongation for Trogamid T® at 20°C (according to Ref. 4).

8.4.3 Fabrication

Granular-form Trogamid can be fabricated directly from the melt with standard commercial equipment; however, it should be predried to avoid surface defects. Fabrication-temperature generally ranges between 250 and

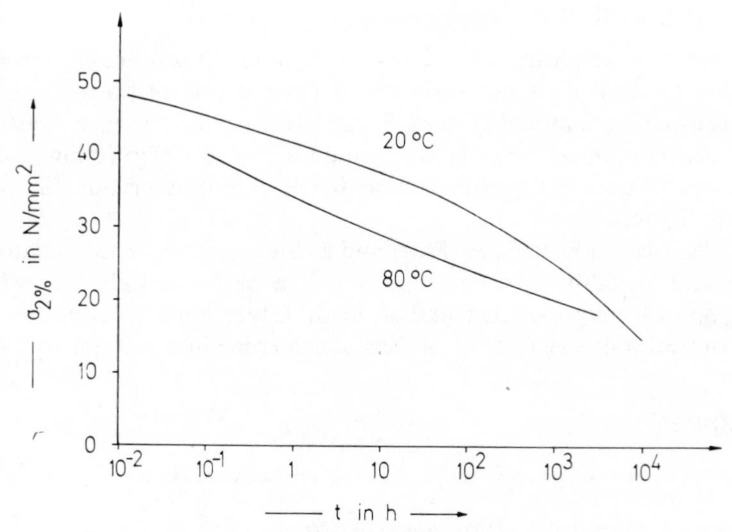

Figure 8-3 Tensile stress at maximum load for Trogamid T® at 2% elongation as a function of loading time at two different temperatures (according to Ref. 4).

320°C,[4] injection-pressures up to 13,000 N/cm². Trogamid has a relatively high melt viscosity and an average shrinkage of about 0.5%. Mold temperatures should be about 70–90°C.

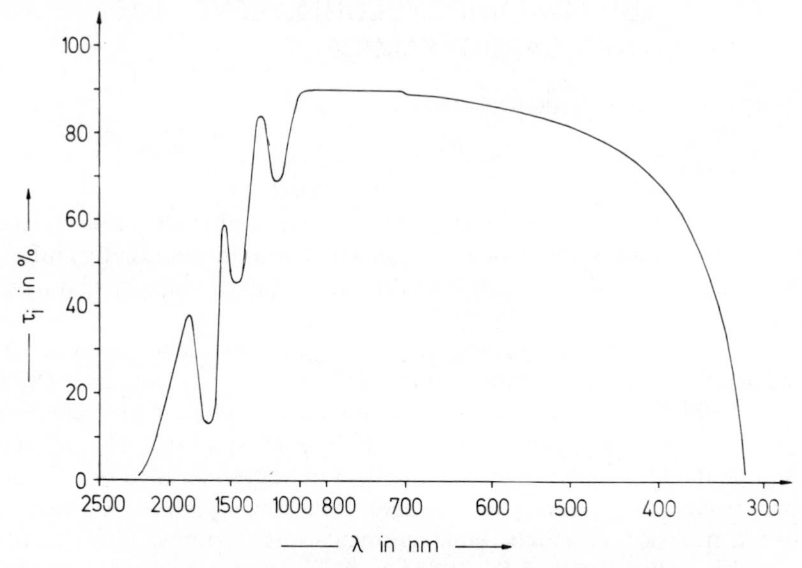

Figure 8-4 Internal transmittance of Trogamid T® as a function of wavelength (according to Ref. 4). Sample thickness, 4 mm.

8.4.4 Application

Trogamid competes with polycarbonate; however, it is more expensive. On the other hand, it does not show the stress-cracking of polycarbonate. A fully transparent material (Figure 8-4), it is, therefore, used as clear pipe fittings and containers for oil sprayers and for a variety of injection moldings with metal inserts. Trogamid is also finished into reservoirs for butane cigarette lighters.

The electrical industry uses Trogamid as high-performance insulators for transparent switchboards, covers and switch parts, as well as for switch knobs, spools, microswitches and so forth. Other areas of application are in the optical industry and for objects which come into contact with foods.

References

1. R. Gabler, H. Müller, G. E. Ashby, E. R. Agouri, H.-R. Meyer and G. Kabes, *Chimia* [*Aarau*], **21**, 65 (1967).
2. H. Doffin, W. Pungs and R. Gabler, *Kunststoffe*, **56**, 542 (1966).
3. G. Bier, "New polycondensation polymers", in N. A. Platzer (Ed.), *Addition and Condensation Polymerization Processes, Adv. Chem. Ser.*, **91**, 612 (1969).
4. J. Schneider, *Kunststoffe*, **64**, 365 (1974).
5. Product information literature of the firm Dynamit Nobel.

8.5 POLYBIS(4-AMINOCYCLOHEXYLENE)-1,10-DECANEDICARBOXYAMIDE

This polymer with the structural element

$$-NH-\langle C_6H_{10}\rangle-CH_2-\langle C_6H_{10}\rangle-NH-OC-(CH_2)_{10}-CO-$$

and the IUPAC-name of poly(iminocyclohexylenemethylenecyclohexylene-imino-1,12-dioxododecamethylene) was brought onto the market by DuPont under the tradename Qiana®. 70% of the cycloalkane groups have a *trans*-configuration.

Starting materials for its synthesis are aniline and butadiene[1] (see Eq. 8.8).

The polymer has a melting temperature of 145°C in the dry state and of 85°C in the wet, which give greater wrinkle resistance and better shape retention. Its 275°C melting point permits processing by melt-spinning. The polymer is insoluble in chlorinated organic solvents and, therefore, is dry-cleanable.

Fibers from this polymer yield silk-like, soft, flowing fabrics for blouses, nightwear, neckties, raincoats, outer garments and curtains, which can be printed with brilliant colors. Spots are easy to remove. A blend with 25% cotton is made for use as underwear.

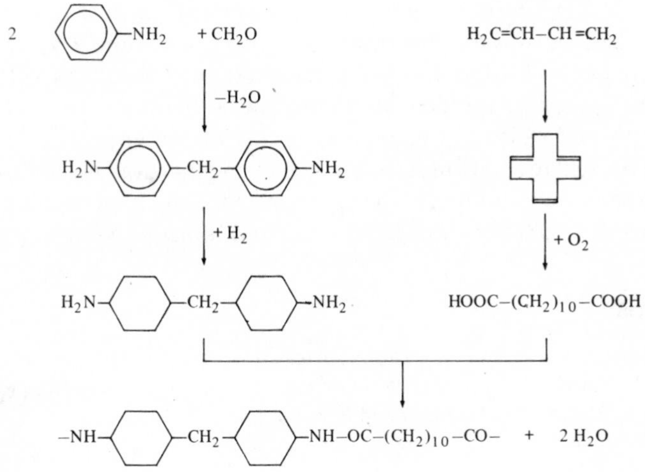

$$2\ C_6H_5\text{–}NH_2 + CH_2O \xrightarrow{-H_2O} H_2N\text{–}C_6H_4\text{–}CH_2\text{–}C_6H_4\text{–}NH_2 \xrightarrow{+H_2} H_2N\text{–}C_6H_{10}\text{–}CH_2\text{–}C_6H_{10}\text{–}NH_2$$

$$H_2C{=}CH\text{–}CH{=}CH_2 \longrightarrow \text{cyclododecatriene} \xrightarrow{+O_2} HOOC\text{–}(CH_2)_{10}\text{–}COOH \tag{8-8}$$

$$\longrightarrow \text{–}NH\text{–}C_6H_{10}\text{–}CH_2\text{–}C_6H_{10}\text{–}NH\text{–}OC\text{–}(CH_2)_{10}\text{–}CO\text{–} + 2\ H_2O$$

References

1. D. W. van Krevelen, *Chimia*, **28**, 504 (1974).

8.6 POLY(NORBORNANAMIDE)

Hoechst AG recently announced a new glass-clear cycloaliphatic polyamide[1] which is apparently a co-polyamide of (a) a mixture of isomeric bisamino-methylnorbornanes (I and II)

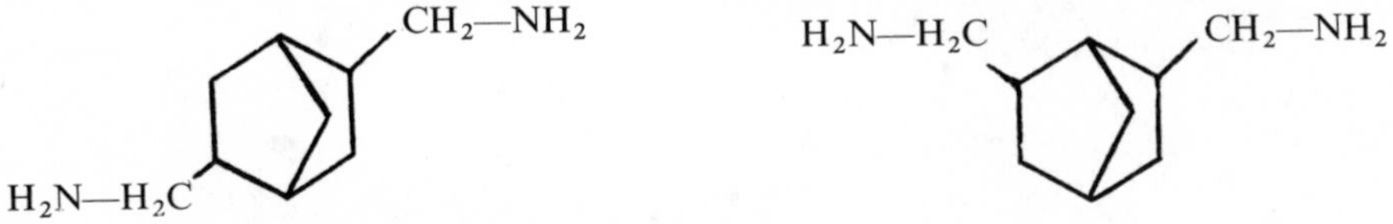

CH_2—NH_2 / H_2N—H_2C (I) H_2N—H_2C / CH_2—NH_2 (II)

I II

with (b) aliphatic or cycloaliphatic dicarboxylic acids with 2–20 carbon atoms or aromatic dicarboxylic acids with 7–20 carbon atoms, as well as (c) other aliphatic, cycloaliphatic or aromatic diamines and (d) amino acids or their lactams.[2] Polymerization proceeds by melt polycondensation. The commercial product is marketed under the tradename Hostamid®; its exact composition has not been disclosed (probably diamines I and II with terephthalic acid and up to 70% ε-caprolactam).

Hostamid® LP 700 is an amorphous polyamide with a density of 1.17 g/cm^3,[3] a glass transition temperature of about 150°C[1] and a Vicat softening temperature of 153°C (dry) and 143°C (in moist air).[3] It softens at 180–210°C and can be injection molded at temperatures between 270 and 310°C.[3] Electrical properties do not change markedly on water absorption of max. 1–1.5%. The surface hardness is high; ball indentation hardnesses of 205 (dry) and 185 N/mm^2 (in moist air) are reported. Tensile strengths at fracture are given as 95 and 91 N/mm^2 for the dry and moist air states.

References

1. Anonym., *Österr. Chem.-Zeitschr.*, **76**(9), 8 (1975).
2. German Pat. 2156723 (November 16, 1971/May 24, 1973); Farbwerke Hoechst AG; Inv.: E. Reske, L. Brinkmann, H. Fischer and F. Rohrscheid.
3. Product information literature, Hoechst AG.

9 Aramid-Fibers

9.1 INTRODUCTION

Aramid-fibers were defined by the U.S. Federal Trade Commission as follows:

A manufactured fiber in which the fiber-forming substance is a long-chain synthetic polyamide in which at least 85 per cent of the amide linkages are attached directly to two aromatic rings.

The first representative of this class was introduced in the 1960's by DuPont under the tradename Nomex®, earlier called HT-1. It is poly(*m*-phenylene-isophthalamide). Besides many advantages, Nomex has one serious disadvantage: its fabrics undergo relatively pronounced shrinkage when exposed to flame. This shrinkage can be tolerated in loose-fitting protective clothing, but not in material for uniforms, usually made from only a single layer of cloth. According to a report Nomex is to be replaced gradually with an HT-4-fiber.[1] It is unclear whether HT-4 is an aramid-fiber or whether it is one of the Kevlar-fibers described in the following section. The properties of HT-4 fibers are listed in Table 9-1.

TABLE 9-1
Properties of Nomex and HT-4 (according to Ref. 2).

Property	Physical unit	Values measured for Nomex	Values measured for HT-4
Titer	tex	0.17	0.11
Tenacity	km	36	34
Ultimate elongation	%	35	6
Modulus	N/mm^2	9100	13,000
Limiting oxygen index (LOI)	%	29	40
Shrinkage[a] after 2 s	%	55	0
after 6 s	%	60	5
Break-up time for the formation of a hole	s	1.2	<60

[a]Under the action of 10.5 $J\ cm^{-2}\ s^{-1}$.

The above Federal Trade Commission definition does not specify whether aramid-fibers may contain other functional groups and whether these must necessarily be bound to the aromatic groups. If such an interpretation proves to be correct, then the poly(amide-hydrazides) (Section 9.3) and the poly-(quinazolinediones) (Section 9.4) also qualify as aramid-fibers.

References

1. G. Pamm and R. A. Hentschel, *Lenzinger Berichte*, **36**, 57 (1974).

9.2 POLY(*p*-PHENYLENETEREPHTHALAMIDE)

9.2.1 Structure and Synthesis

Poly(*p*-phenyleneterephthalamide) or PPDT has the repeat unit

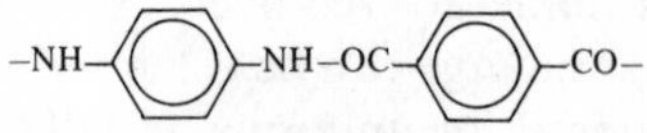

$$-NH-C_6H_4-NH-OC-C_6H_4-CO-$$

DuPont has produced PPDT since 1973, first under the tradename Fiber B®, later as Kevlar®.[1-4] It is synthesized by the treatment of *p*-phenylene-diamine with terephthalyl chloride in hexamethylphosphoramide/*N*-methyl-pyrrolidone (2:1) at $-10°C$:

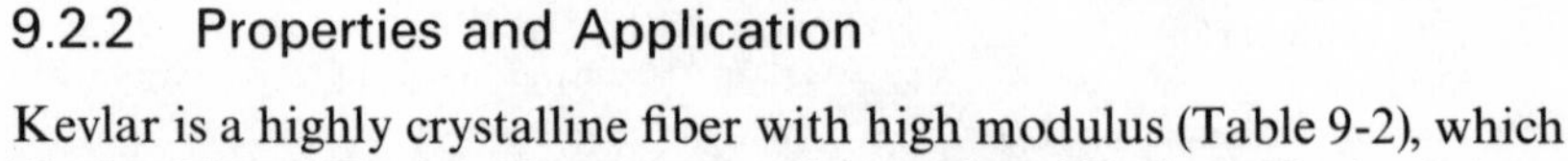

$$H_2N-C_6H_4-NH_2 + ClOC-C_6H_4-COCl \xrightarrow{-2\,HCl} -\!\!\!\!(HN-C_6H_4-NH-OC-C_6H_4-CO)\!\!\!\!- \qquad (9\text{-}1)$$

Kevlar 29 and 49 are also aramid-fibers. Their chemical structure is not published. Earlier Kevlar 49 was marketed under the tradename PRD-49.

9.2.2 Properties and Application

Kevlar is a highly crystalline fiber with high modulus (Table 9-2), which like Kevlar 29 and Kevlar 49 competes with steel and E-glass. Kevlar is intended for use as tire cord in radial tires.

Kevlar 29 and Kevlar 49 are self-extinguishing when the flame is removed. The polymers char but do not melt. Both types are compatible with most epoxide-, unsaturated polyester-, phenol, and polyimide-resins, consequently they are used to reinforce these polymers. Possible areas of application are in the aviation-, electrical- and electronic industries and in boat hulls and sporting goods. Kevlar 49 is not dyeable and has a negative coefficient of thermal expansion of $-2 \times 10^{-6}\ K^{-1}$.

TABLE 9-2
Properties of Kevlar polymers.

Property	Physical unit	Values measured for Kevlar Refs. 5 and 6	Kevlar 29 Ref. 7	Kevlar 49 Ref. 7
Density	g/cm^3	1.45	1.44	1.45
Tensile strength	N/mm^2	2500	2800	3700
Modulus	N/mm^2	58,000 −150,000	630,000	134,000
Ultimate elongation	%	2	3–4	2.8
Tenacity	km		198	
Specific modulus	$N\ cm\ g^{-1}$		62,000	
Useful temperature range	°C		−250 to +260	−200 to +200

References

1. U.S. Patent 3671542 (June 12, 1968); DuPont; inv.: S. L. Kwolek.
2. German Patent 1810426 (February 12, 1970); DuPont; inv.: S. L. Kwolek; *C. A.*, **72**, 112676t (1970).
3. German Patent 1929694 (September 3, 1970); DuPont; inv.: S. L. Kwolek, *C. A.*, **83**, 121469a (1970).
4. German Patent 1929713 (February 12, 1970); DuPont; Inv.: S. L. Kwolek, *C. A.*, **72**, 122796w (1970).
5. G. Hinrichsen, R.Miessen and M. Reichardt, *Angew. Makromol. Chem.*, **40/41**, 239 (1974).
6. D. W. van Krevelen, *Chimia* [*Aarau*], **28**, 505 (1974).
7. Product information literature of DuPont.

9.3 POLY MIDE

9.3.1 Structure and Synthesis

The Monsanto Company produces a series of high modulus fibers under the designation X-500, apparently comprised of both aromatic polyamides as well as aromatic poly(amide-hydrazides). The primary member seems to be an "ordered" poly(amide-hydrazide) designated as PABH-T X-500:

$$-NH-C_6H_4-CO-NHNH-OC-C_6H_4-CO-$$

The properties of this polymer were described in detail in papers at the April 1972 American Chemical Society Meeting and were published as summaries,[1] original articles,[2] and a book.[3]

As judged from these publications, two more polymers appear interesting for Monsanto, namely another poly(amide-hydrazide)

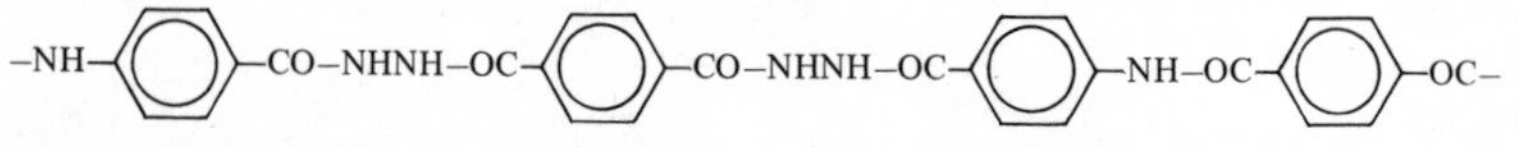

$$-NH-C_6H_4-CO-NHNH-OC-C_6H_4-CO-NHNH-OC-C_6H_4-NH-OC-C_6H_4-OC-$$

and the terephthalamide of 4,4′-diaminobenzanilide

$$-NH-C_6H_4-CO-NH-C_6H_4-NH-OC-C_6H_4-CO-$$

The publications also contain additional data for a series of aromatic polyamides.

PABH-T X-500 is formed by the reaction of *p*-aminobenzhydrazide with terephthalyl chloride

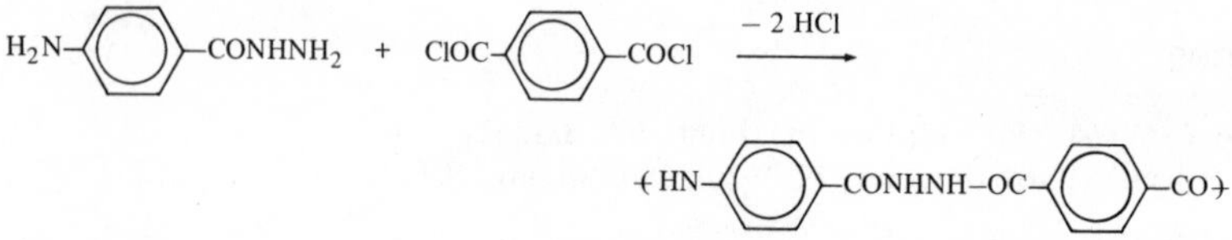

$$H_2N-C_6H_4-CONHNH_2 + ClOC-C_6H_4-COCl \xrightarrow{-2\,HCl} \left[HN-C_6H_4-CONHNH-OC-C_6H_4-CO\right] \qquad (9\text{–}2)$$

TABLE 9-3

Properties of various poly(amide-hydrazides) (according to Ref. 5).

Property	Physical unit	Values measured for PABH-T(G)	PABH-T(T)	PABH-T(E)
Density	g/cm^3	1.47	1.44	
Titer	tex	0.44–0.88	0.55–1.1	
Degree of crystallinity	%	85–93	30	
Birefringence	1	0.57	0.3	
Decomposition temperature	°C	525	390	
Tensile strength	N/mm^2	~2100	~1100	~700
Ultimate elongation	%	3–4	15–20	20–40
Modulus	N/mm^2	~95,000	~35,000	~15,000
Creep modulus				
(1 min at 387 N/mm^2)	N/mm^2	76,000		
(1 min at 972 N/mm^2)	N/mm^2	60,000		
Knot strength	N/mm^2	200	960	
Flex life	1		~90	

Because of the differing reactivities of the amino- and the hydrazide-groups, a largely ordered structure is formed. Occasional "false" reactions can appear, and as a result the Monsanto researchers prefer to speak of "partially ordered" poly(amide-hydrazides). Differentiation is made between three types of PABH-T X-500:

PABH-T(G) X-500: extensibility and modulus like glass
PABH-T(T) X-500: extensibility like a high strength fiber, usable as tire cord
PABH-T(E) X-500: special type with higher extensibility.

TABLE 9-4
Modulus and specific modulus (=modulus/density) of various fibers (according to data in Refs. 5 and 6).

			Properties	
Fiber	Density g/cm^3	Fiber diameter μm	Modulus N/mm^2	Specific modulus N cm g^{-1}
Poly(amide-hydrazide) PABH-T(G)	1.47	25	106,000	720
Graphite (Modmor I)	1.99	7.5	420,000	2100
Steel (stretched)	7.75	75	210,000	270
E-glass	2.55	5–10	70,000	270
Boron	2.63	100	350,000	1300
Aluminium	2.68		70,000	260
Beryllium	1.83		320,000	1700
Poly(amide-hydrazide) PABH-T(T)	1.44		350,000	240
Rayon	1.53		16,000	100
Poly(ethylene terephthalate) (high tensile)	1.39		10,000	72
Polyamide 6 (high tensile)	1.14		4500	40
Poly(amide-hydrazide) PABH-T(E)	(1.46)		15,000	100
Poly(ethylene terephthalate) (Textile fiber)	1.385		5600	40
Polyamide 6 (Textile fiber)	1.12		3500	30

9.3.2 Properties and Application

Poly(amide-hydrazides) PABH-T X-500 (I) are pale-straw-colored polymers with molecular weights of 25,000–65,000 g/mol.[4] A few published properties are listed in Table 9-3. This table shows that these fibers have very high moduli. The so-called specific moduli (=modulus/density) are more than double those of steel wire and glass fibers (Table 9-4) and considerably higher than other organic polymers, with the exception of graphite fiber.

The properties of the poly(amide-hydrazide) II are given in Table 9-5. As is apparent from the sparse published data, the modulus is strongly influenced by the method of fabrication.

TABLE 9-5
Properties of poly(amide-hydrazide) II.

Property	Physical unit	Values measured: not hot-stretched	hot-stretched
Density	g/cm^3	1.44	1.47
Titer	tex	0.40	0.30
Tenacity	km	74	97
Ultimate elongation	%	9.4	2.9
Modulus	N/mm^2	38,000	67,000

The fibers from these polymers all have very high moduli, low densities, limited moisture sensitivity and high abrasion resistance. Therefore, they are used as tire cords and as reinforcing agents for plastics.

References

1. Series of Abstracts in *ACS Polymer Preprints*, **13**, 619–623 (1972).
2. Series of articles in *J. Macromol. Sci.* [*Chem.*], **A7**, 3–348 (1973).
3. W. B. Black and J. Preston (Eds.) *High-Modulus Wholly Aromatic Fibers*, M. Dekker, New New York, 1973.
4. J. Burke, *J. Macromol. Sci.* [*Chem.*], **A7**, 187 (1973).
5. W. B. Black, J. Preston, H. S. Morgan, G. Rauman and M. R. Lilyquist, *J. Macromol. Sci.* [*Chem.*], **A7**, 137 (1973).
6. M. R. Lilyquist, R. E. DeBrunner and J. K. Fincke, *J. Macromol. Sci.* [*Chem.*], **A7**, 203 (1973).
7. J. Preston, W. B. Black and W. L. Hofferbert, Jr., *J. Macromol. Sci.* [*Chem.*], **A7**, 45 (1973).

9.4 POLY(QUINAZOLINEDIONE)

9.4.1 Structure and Synthesis

Aromatic polyamides with the characteristic quinazolinedione-group in the structural element

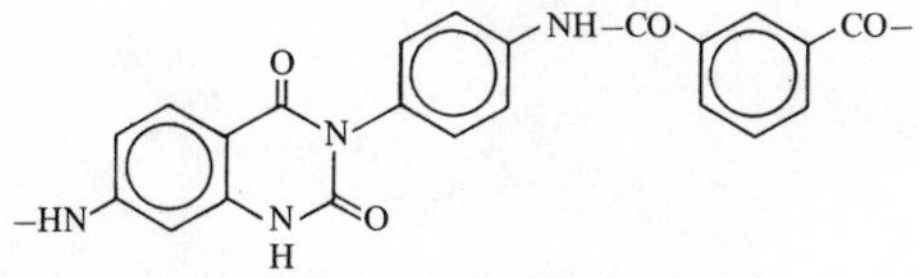

are produced by the firm Bayer under the tradename AFT-2000. The quinazolinedione-group is synthesized as follows:[1-3]

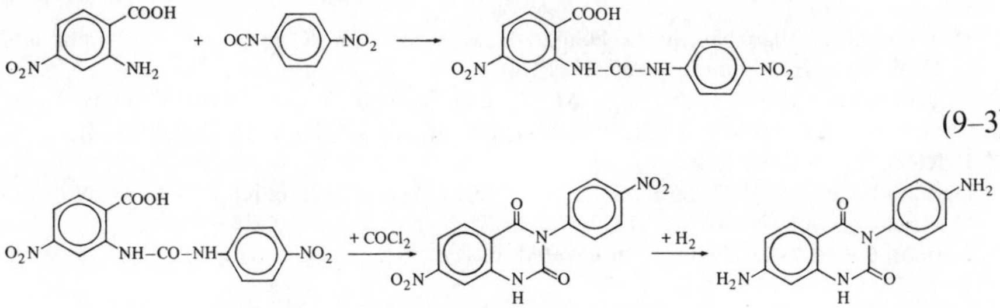

(9–3)

The resulting diamine is dissolved in *N*-methylpyrrolidone or dimethylacetamide and then polycondensed with isophthalyl chloride to give AFT-2000:

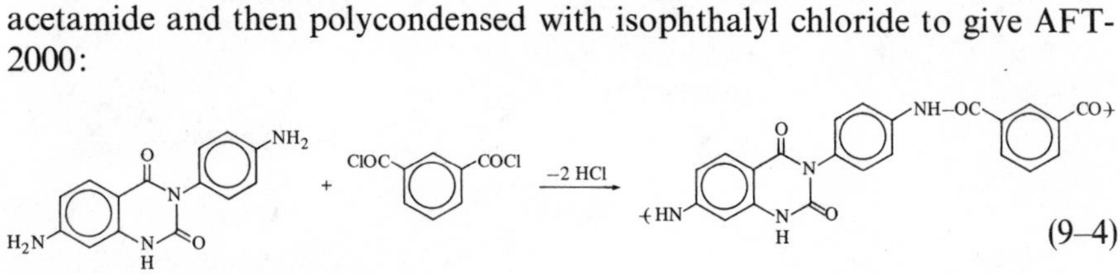

(9–4)

The viscous solution can then be directly dry- or wet-spun.[4,5]

9.4.2 Properties

AFT-2000 is X-ray amorphous, even after stretching.[4,6] The glass temperature has not yet been reported. The fiber still has 90% tensile strength after 100 h at 200°C.[6] Heating at 500°C for a short time reduces the strength by only about 20%.

Fiber cannot be made by melt spinning because the softening temperature lies above the decomposition temperature.[4] Dry-spinning gives smooth fibers with dumbbell-shaped cross sections, wet spinning laterally fibrillated fibers with circular cross sections.[6] The fibers can absorb 9.5 to 11% water.[7] The tensile stress is reported as 0.5 N/tex, the ultimate elongation as 12–15% and the modulus as 100 N/mm^2.[6] The fibers can be dyed with basic-, acid-, metal complex- and dispersion dyes.[4] The limiting oxygen index is 36–40%.[4]

Use of this hygroscopic, thermally stable and flame-resistant fiber in fabric for protective clothing and as filter felts for hot-gas filtration is foreseen.[8,9]

References

1. German Patent 1720686 (July 15, 1971); Bayer AG.
2. H. E. Küntzel, G. D. Wolf, F. Bentz, G. Blankenstein and G. E. Nischk, *Makromol. Chem.*, **130**, 103 (1969).
3. H. E. Küntzel, F. Bentz, G. D. Wolf, G. Blankenstein and G. E. Nischk, *Makromol. Chem.*, **138**, 223 (1970).
4. B. v. Falkai, G. Blankenstein, G. Hentze, H. E. Küntzel, P. Kleinschmidt, G. Nischk and G. D. Wolf, *Lenzinger Berichte*, **36**, 48 (1974).
5. Belgian Patent 718033 (1968); Bayer AG; inv.: G. D. Wolf, W. Giessler and F. Bentz.
6. G. Hinrichsen, R. Miessen, M. Reichardt, *Angew. Makromol. Chem.*, **40/41**, 239 (1974).
7. P. Kleinschmidt, cited in Ref. 6.
8. German Patent 2103877 (January 28, 1971); Bayer AG; inv.: H. E. Küntzel, G. D. Wolf and G. Nischk; *C. A.*, **77**, 140134m (1972).
9. German Patent 1802079 (October 9, 1968); Bayer AG; inv.: H. E. Küntzel and G. D. Wolf; *C. A.*, **73**, 45537d (1970).

10 Polyimides

10.1 STRUCTURE

Polyimides have the characteristic functional group

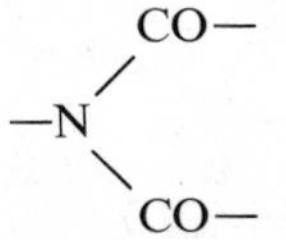

$$-N\begin{matrix} \diagup CO- \\ \diagdown CO- \end{matrix}$$

This imide group can be bound or contained in a number of base units, and the polymer structure can be obtained either by polycondensation or by polymerization. The "true" polyimides are to be distinguished from polyamideimides and polyesterimides, also polycondensates:

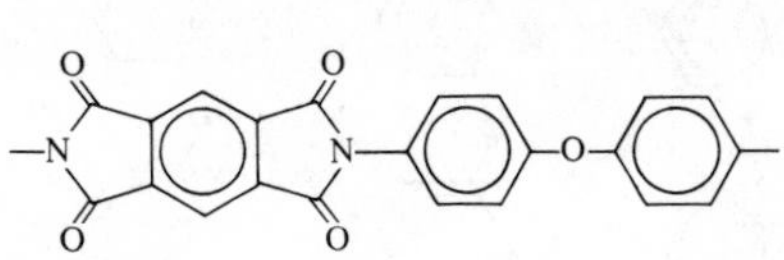

polyimide I

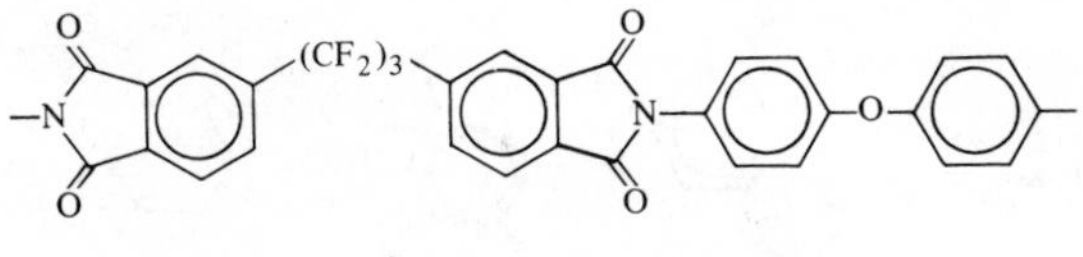

polyimide Ia

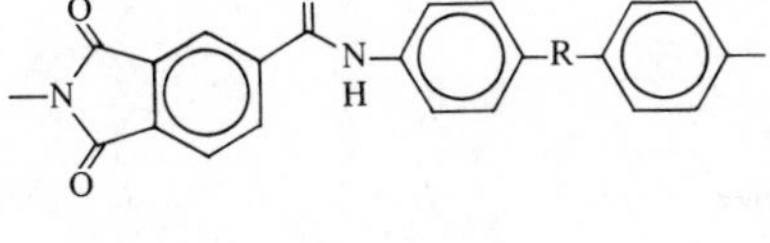

polyamideimide II

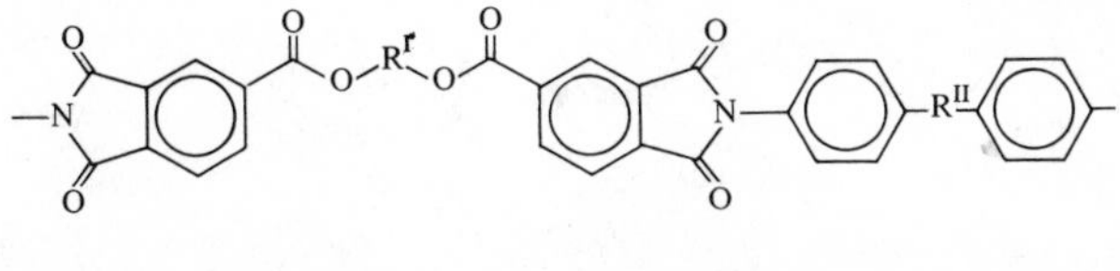

polyesterimide III

Polyimides, polyamideimides, and polyesterimides cure by polycondensation, whereas polybismaleimides (IV) and the polyimides (V–VI) cure by addition polymerization.

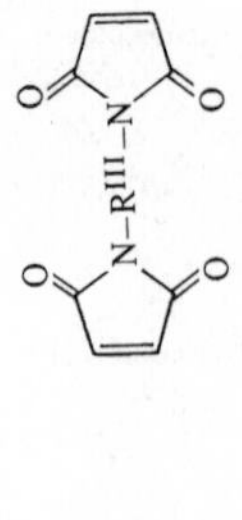

poly(bismaleinimide) IV

R^{III} =

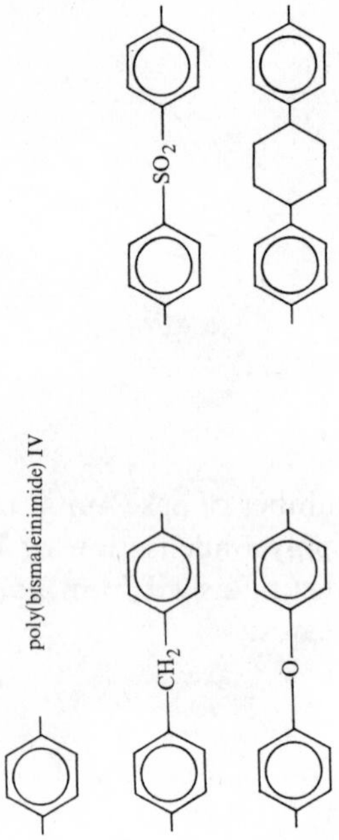

SO_2

CH_2

O

Va $HC{\equiv}C$ … $C{\equiv}CH$ (HR 600)

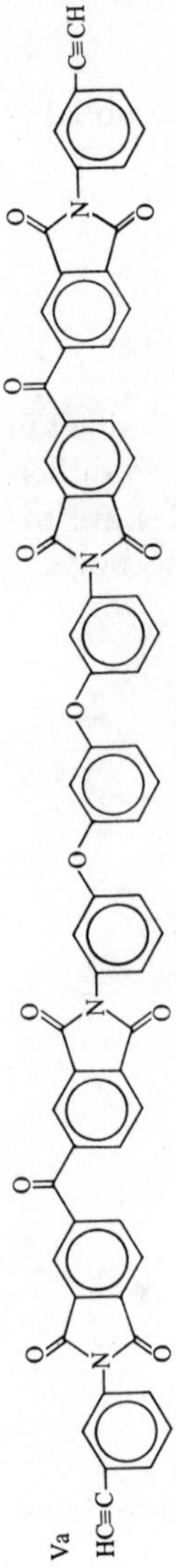

Vb $HC{\equiv}C$ … $C{\equiv}CH$ (HR 650)

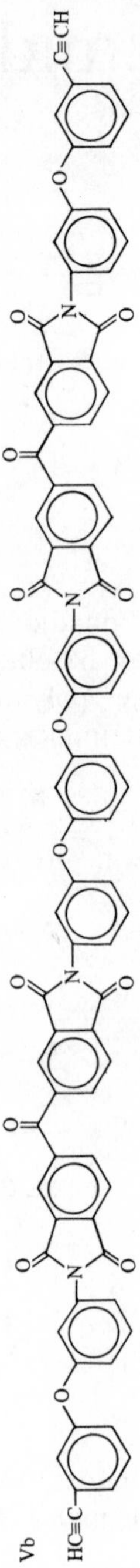

Vc $HC{\equiv}C$ … $C{\equiv}CH$ (HR 700)

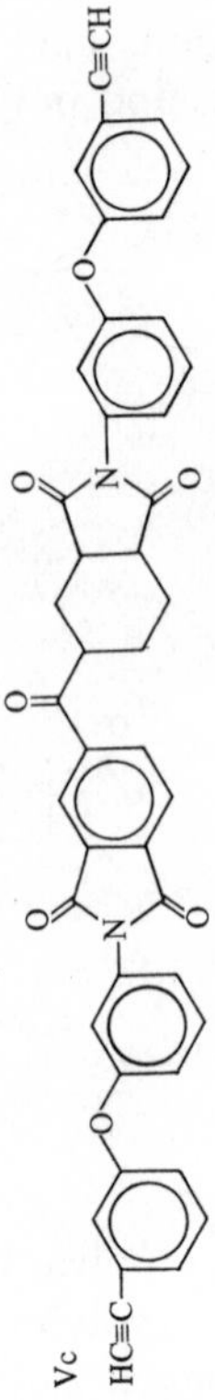

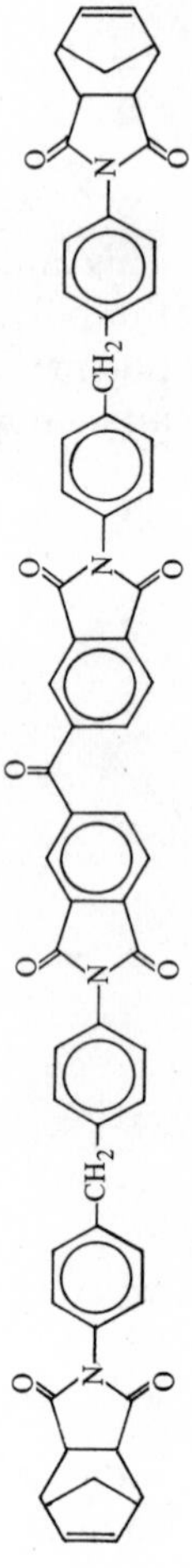

CH_2 … CH_2 (P 13 N)

Type I polyimides were first marketed over 10 years ago for wire coating. These were followed by the polyamideimides (II) and the polyesterimides (III), and most recently by the polymerizable types IV–VI. The areas of application have also greatly expanded. Besides wire lacquer, these now include dry powder, compounds with glass fibers, graphite, molybdenum disulfide, asbestos fibers, poly(tetrafluoroethylene) and diamond dust, semi-finished articles, prepolymerized prepregs, laminates, honeycombs,

TABLE 10-1
Commercially available polyimides: Tp = thermoplastic; Ts = thermoset; R = resin; C = compound prepreg; M = molded article; F = film; [a] = unmeltable.

Trade name	Firm	Type	Poly-condensate	Poly-merizate	Form in which supplied
2225	Bayer	I			
AI-630	Amoco	II	Tp		R
Amanim	Westinghouse				
Durette X-400	Monsanto				
E 3520	Herberts	III			
Feurlon	Bemol		Ts		M
Germon	General El.	IV			
Icdal Ti 40	Dynamit Nobel	III			
Imipex	General El.	III			
—	Hughes	V			
—	Hexcel		Ts		C
Kapton	DuPont	I	T[a]		F
Kermel	Rhône-Poulenc	II			
Kerimid	Rhône-Poulenc			Ts	R
Kinel M 33	Rhône-Poulenc	IV		Ts	R
Meldin	Dixon		------Ts and Tp------		M
P 13 N	Ciba-Geigy	VI		Ts	R
Polyimidal	Raychem		Tp		R
Polyimide 2080	Upjohn		------Tp------		R
Pyralin	DuPont	I	Tp[a]		R, C
Pyre ML	DuPont	I	Tp[a]		R
QX 13	ICI				
Rhodeftal	Rhône-Poulenc			Ts	R
Skybond	Monsanto		Ts		R
Sparmon	Sparta Mfg.			Ts	M
Terebec	Dr. Beck	III			
Torlon	Amoco	II	Tp		R
Tritherm 981	P.D. George		Tp		R
Vespel	DuPont	I	Tp[a]		M
XP 182	Amer. Cyanamide				
XWE 810	Schenectady Chem.			Tp	R
YF-Series	LNP			Tp	C

adhesives and foams. The tradenames of the old and new products, their chemical structure (in so far as is known), the suppliers, and the form in which supplied are given in Table 10-1. The synthesis and properties of polyimides have been described in several books and review articles.[2-5]

10.2 SYNTHESIS

The first commercial polyimide (I) was produced through the polycondensation of pyromellitic anhydride with 4,4-diaminodiphenyl ether. The first step gives a polyamide carboxylic acid

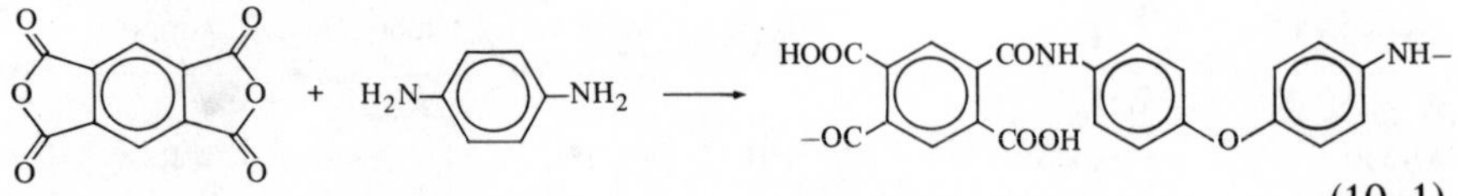

$$+ \; H_2N\text{–}\langle\bigcirc\rangle\text{–}NH_2 \longrightarrow \; HOOC,\ \text{–OC},\ CONH,\ COOH,\ O,\ NH\text{–} \qquad (10\text{–}1)$$

which is then converted by dehydration and ring closure in the second step to the polyimide

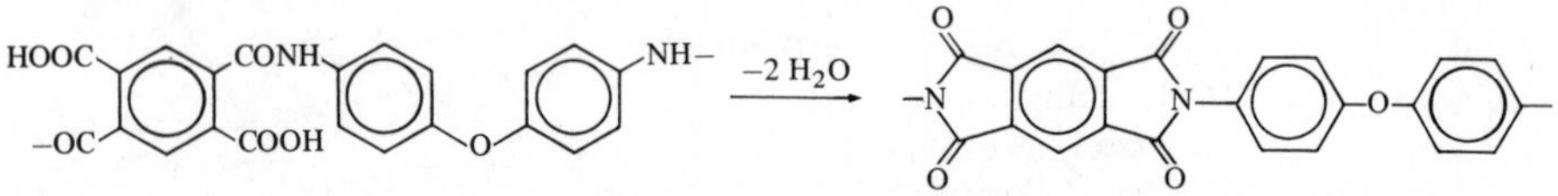

$$HOOC,\ \text{–OC},\ CONH,\ COOH,\ O,\ NH\text{–} \xrightarrow{-2\,H_2O} \text{–N},\ N\text{–},\ O \qquad (10\text{–}2)$$

The polycondensation in the second step is both intramolecular and intermolecular. Cross-linking is the result; consequently, product forming must be executed during the second step. Also, during the second step water is released, vaporized, and incorporated into the solidifying mass. The resulting voids reduce the mechanical properties and lower the thermal stability.

Use of trimellitic anhydride in place of pyromellitic anhydride leads to polyamideimides

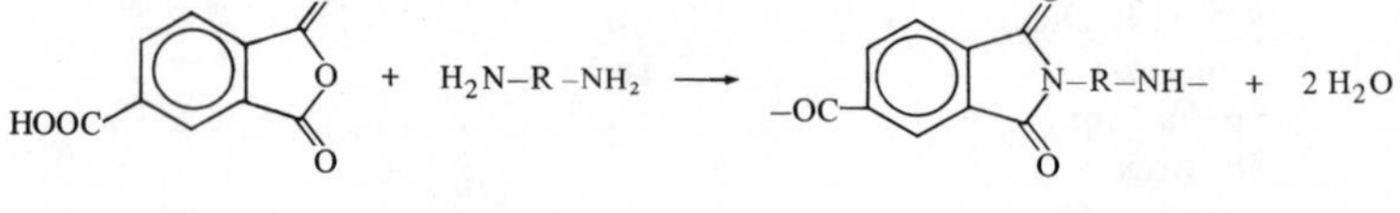

$$HOOC \; + \; H_2N\text{–}R\text{–}NH_2 \longrightarrow \text{–OC} \; N\text{–}R\text{–}NH\text{–} \; + \; 2\,H_2O \qquad (10\text{–}3)$$

Reactions (10–1) to (10–3) require high purity starting materials. Because of the difficult purification and the high basicity of the amines, substitution with "capped" amines has been suggested (see Ref. 6):

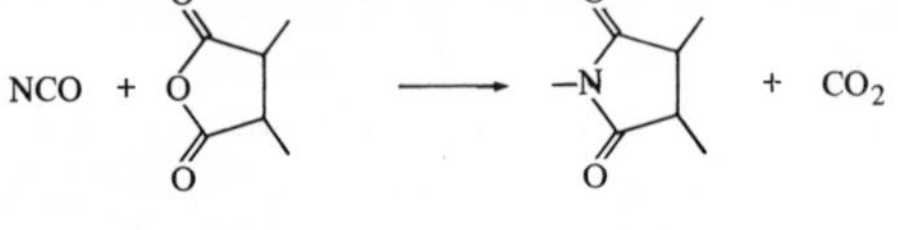

$$NCO \; + \; O \longrightarrow \text{–N} \; + \; CO_2 \qquad (10\text{–}4)$$

$$-NH-COOR + (\text{anhydride}) \longrightarrow -N(\text{imide}) + CO_2 + ROH \tag{10–5}$$

$$-NH-CO-NH + (\text{anhydride}) \longrightarrow -N(\text{imide}) + CO_2 + -NH_2 \tag{10–6}$$

Reaction (10–4) or (10–5) is used in the commercial synthesis of Kermel-fibers. Trimellitic anhydride is treated with a diisocyanate or a diurethane to give polyamideimides:[7]

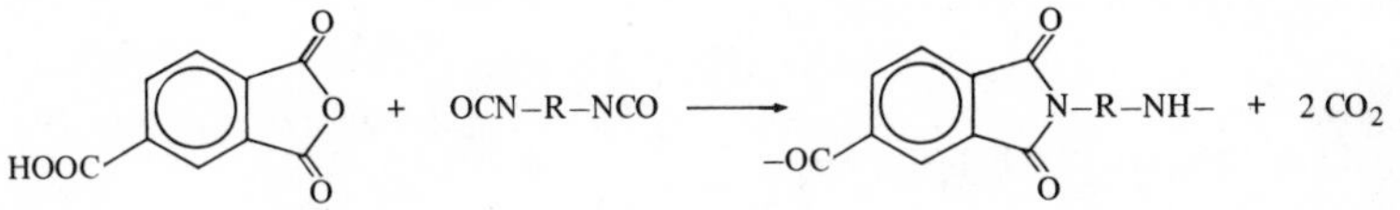

$$HOOC-(\text{trimellitic anhydride}) + OCN-R-NCO \longrightarrow -OC-(\text{phthalimide})N-R-NH- + 2\,CO_2 \tag{10–7}$$

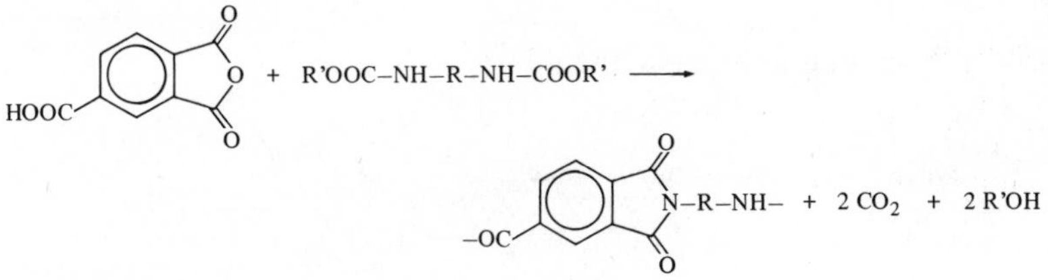

$$HOOC-(\text{trimellitic anhydride}) + R'OOC-NH-R-NH-COOR' \longrightarrow -OC-(\text{phthalimide})N-R-NH- + 2\,CO_2 + 2\,R'OH \tag{10–8}$$

Upjohn's new one-step procedure goes directly from the isocyanate of trimellitic acid

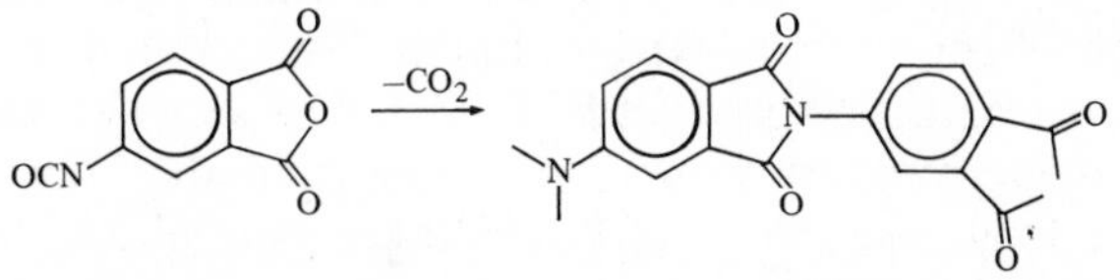

$$OCN-(\text{phthalic anhydride}) \xrightarrow{-CO_2} \text{polyimide} \tag{10–9}$$

Addition-reactions on bismaleimides do not produce volatile, low molecular weight by-products. Bismaleimides are synthesized in the reaction of maleic anhydride with diamines:

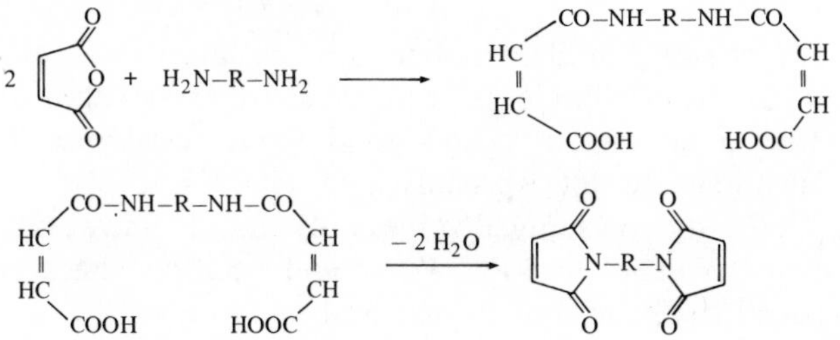

$$2\,(\text{maleic anhydride}) + H_2N-R-NH_2 \longrightarrow HOOC-CH=CH-CO-NH-R-NH-CO-CH=CH-COOH$$

$$HOOC-CH=CH-CO-NH-R-NH-CO-CH=CH-COOH \xrightarrow{-2\,H_2O} (\text{maleimide})N-R-N(\text{maleimide}) \tag{10–10}$$

A variety of bifunctional compounds add to bismaleimides, namely amines[8,9] to give poly(aspartimides)

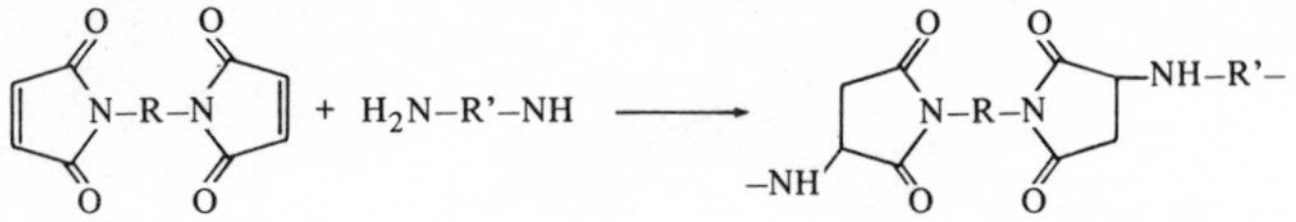

$$\text{bismaleimide (N-R-N)} + H_2N{-}R'{-}NH \longrightarrow \text{-NH-[succinimide]N-R-N[succinimide]-NH-R'-} \quad (10\text{–}11)$$

sulfides[10]

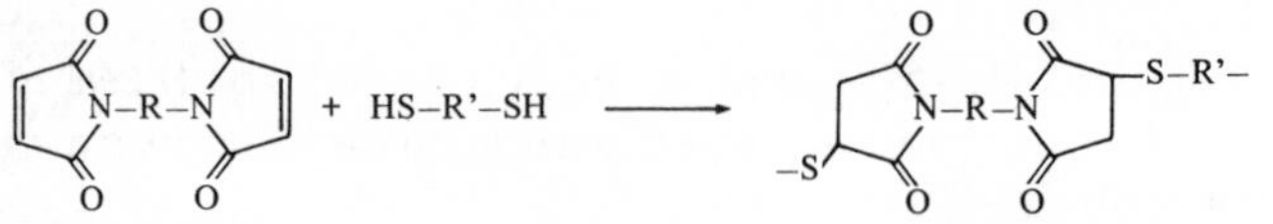

$$\text{bismaleimide (N-R-N)} + HS{-}R'{-}SH \longrightarrow \text{-S-[succinimide]N-R-N[succinimide]-S-R'-} \quad (10\text{–}12)$$

aldoximes[11]

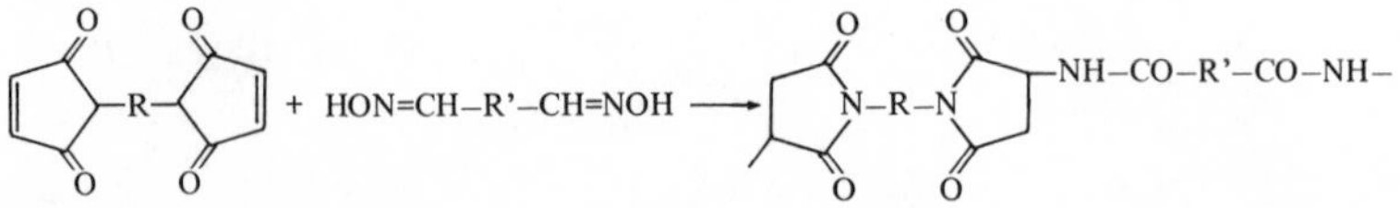

$$\text{bismaleimide (-R-)} + HON{=}CH{-}R'{-}CH{=}NOH \longrightarrow \text{[succinimide]N-R-N[succinimide]-NH-CO-R'-CO-NH-} \quad (10\text{–}13)$$

and to dichlorobismaleimides, also phenols[12]

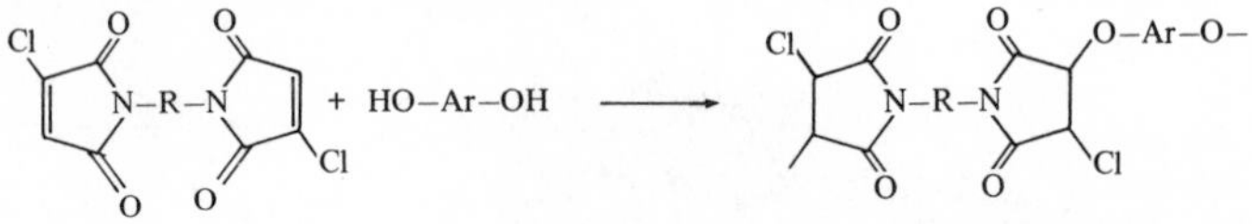

$$\text{Cl-bismaleimide (N-R-N)-Cl} + HO{-}Ar{-}OH \longrightarrow \text{Cl-[succinimide]N-R-N[succinimide](Cl)-O-Ar-O-} \quad (10\text{–}14)$$

The addition of equimolar amounts of diamines apparently does not give usable commercial products.[8] Reaction of nonequimolar amounts of diamines at high temperatures leads to both the addition of the amines and to polymerization of the carbon–carbon double bond. Polymers which have good properties result. The product P13N of Ciba-Geigy(VI) cures by polymerization of the terminal carbon–carbon double bonds.[18] The polymerization of the acetylenic end-groups of Va–Vc takes place above their softening temperature of about 180°C, apparently *via* trimerization.[1]

10.3 UNFILLED POLYIMIDES

The properties of some unfilled polyimides are given in Table 10-2. The data show that all the polyimides have high short-term strength, and thermal stability, a low creep tendency and good flame resistance. The impact strengths with notch are not especially high. Heating to 260°C in an inert atmosphere lowers the mechanical values only slightly. However, heating in air causes them to drop 30–40%. Wear and traction characteristics are excellent, especially solvent-free on steel and cast iron.

TABLE 10-2
Properties of unfilled polyimides (according to Refs. 1, 5, 13 and 17).

Property	Physical unit	Values measured for					
		DuPont Vespel	Amoco Torlon 2000	Upjohn 2080	Ciba-Geigy P 13 N	Rhône-Poulenc Kinel M 33	Hughes HR 600
Density	g/cm^3	1.43	1.41	1.4	1.33	1.3	
Coefficient of linear thermal expansion	$10^5 K^{-1}$	5–6	3.4	5.0			
Heat distortion temperature	°C	357	282	270–280			
Tensile strength							
at 23 °C	N/mm^2	76–100	93	120	50		
at 260–300 °C	N/mm^2	33–43	62	30	39		
Ultimate elongation	%		2.5				
Flexural strength							
at 23 °C	N/mm^2	107–128	164	200	69–83	137	169
at 260–300 °C	N/mm^2	58–72	100	35	41–55	49	
Flexural modulus							
at 23 °C	N/mm^2	3030–3170	4900	3310	3170–3380	3730	4650
at 260–300 °C	N/mm^2	1790–1900	3100	1100	2070–2270	2750	
Compressive strength							
at 23 °C	N/mm^2	253–310	>245	206	258		218
at 260–300 °C	N/mm^2	128–133					
Shear strength	N/mm^2		58–94	130			
Impact strength	N/mm^2					24	
Impact strength with notch	J/cm	2.4–4.8	2.4	2.4			
Flex fatigue limit (10^7 Hz)	N/mm^2	45					

TABLE 10-2 ; *cont.*

Property	Physical unit	Values measured for					
		DuPont Vespel	Amoco Torlon 2000	Upjohn 2080	Ciba-Geigy P 13 N	Rhône-Poulenc Kinel M 33	Hughes HR 600
Deformation under load (13.8 N/mm^2 at 60 °C)	%	0.14					
Hardness (Rockwell)		45–58 E	104 E	99 E	91 D		
Stat. coefficient of friction		0.35					
LOI	%			44	32		
Dielectric constant (10^5 Hz)	1	3.4	3.7	3.4	4.3	3.5	3.4
Dissipation factor (10^5 Hz)	1	0.0052	0.001	0.0055	0.006	0.003	0.01
Dielectric strength	kV/cm	220	169		147		
Volume resistivity	Ω cm	10^{16}–10^{17}	3×10^{14}		1.6×10^{16}	10^{14}	
Surface resistivity	Ω	10^{15}–10^{16}	$>10^{15}$		4×10^{14}		
Arc resistance	s	435			83		
Water absorption (24 h)	%	0.32	0.28	0.6	0.4		
equilibrium	%	3					

Polyimides absorb only minor amounts of water which causes only minor changes in dimensions. These are reversible by drying. Polyimides are not stable to hydrolysis; they form cracks in water or steam at temperatures above 100°C. Organic solvents have only a minor effect on the mechanical properties. Polyimide surfaces can even be cleaned with halogenated hydrocarbons. Concentrated mineral acids and bases do attack polyimides. Polyimides stop burning when the flame source is removed. They char and release only a small amount of fumes, but they do not melt.

Molded and machined parts out of polyimides are used in piston rings, bearings, sealants, valve seats, disk breaks, electrical junctions, spools, jet engines and nuclear reactor fittings. Polyimide films are used in cable and wire insulation as supports for printed circuits and as keyway insulation.

10.4 REINFORCED POLYIMIDES

Reinforcement of polyimides with glass fibers, graphite and the like considerably reduces the coefficient of thermal expansion and the creep tendency. This improves the already good mechanical properties (Table 10-3 and 10-4).

Reinforced polyimides are used in the electrical, electronics, and the aircraft industries as plugs, sunk keys, insulating clamps, sheathings, parts for jet engines, fire walls, leading edges of wings, radomes and so forth.

Polyimides reinforced with glass-fiber fabrics are suitable for printed circuits, turbine blades and airplane noses, honey-comb combinations for compressor structural elements and jet-engine parts. Extraordinarily high flexural strengths can be achieved (Table 10-5) which decrease only slowly with time even at 200°C (Figure 10-1). The exact flexural strengths which can be obtained naturally depend on the chemical structure of the particular polyimide. This is shown in Figure 10-2 for the polyimides of various manufacturers.

10.5 FOAMS

Monsanto produces a series of foamable polyimides under the tradename Skybond (previously: RI-7271). Type 01 (with an initial bulk density of about 1 $lb/ft^3 = 0.016$ g/cm^3) foams to about twenty times its original volume and gives a flexible cell structure. Types 06, 12 and 18 (with bulk densities of about 0.1, 0.2 and 0.3 g/cm^3) give rigid foams with no change in dimensions.

Heating type 01 to 175°C yields a red, brittle foam which cures to a yellow, tough foam within minutes at 300°C. Further curing for 4 to 16 h at 300°C is recommended. Types 06, 12 and 18 are foamed directly on heating to

TABLE 10-3

Properties of glass-fiber reinforced polyimides (according to Refs. 5, 8 and 14).

Property	Physical unit	Values measured for: DuPont Vespel	Rhône-Poulenc Kinel 5504	Ciba-Geigy P 13 N	General Electric 3010
Density	g/cm^3	1.75	1.9	1.83	1.90
Coefficient of linear thermal expansion	10^5 K^{-1}	1.8–3.8	1.5		1.4
Heat distortion temperature (at 1.84 N/mm^2)	°C		349		349
Tensile strength					
at 23 °C	N/mm^2	26–49	186	99	197
at 260 °C	N/mm^2		157	64	169
Flexural strength					
at 23 °C	N/mm^2	41–83	343	207–276	394
at 260 °C	N/mm^2		245	172–207	317
Flexural modulus					
at 23 °C	N/mm^2	4300–6900	24,500	13,800–16,500	27,000
at 260 °C	N/mm^2		17,700	11,700–13,100	20,800
Compressive strength					
at 23 °C	N/mm^2		230	225	
at 260 °C	N/mm^2			150	
Impact strength with notch	J/cm		59		9.1
Flex fatigue limit at 70.4 N/mm^2	—				2,200,000
Hardness (Rockwell)	—	44–56 E	125 M	95 D	114 E
Water absorption (24 h)	%		0.2	1.8	0.2
Dielectric constant	1		4.7	6.0	4.8
Dissipation factor	1		0.003	0.006	0.0034
Dielectric strength	kV/cm		220	141	197
Volume resistivity	Ω cm		9×10^{15}	5×10^{15}	
Surface resistivity	Ω			2×10^{14}	
Arc resistance	s			182	50–180

TABLE 10-4
Properties of polyimides reinforced with graphite fiber.

Property	Physical unit	Values measured for				
		DuPont Vespel	Amoco Torlon	Rhône-Poulenc Kinel	Upjohn 2800	Hughes HR 600 C
Density	g/cm^3	1.51	1.45	1.45		
Coefficient of linear thermal expansion	$10^5 K^{-1}$	4.1–6.5		1.9		
Heat distortion temperature at 1.82 N/mm^2	°C	357	296			
Tensile strength at 23 °C	N/mm^2	45–80	90	39	74	
at 260–300 °C	N/mm^2	30–41	36	25		
Modulus	N/mm^2				1870	
Flexural strength at 23 °C	N/mm^2	80–125	152	88	96	560
at 260–300 °C	N/mm^2	45–70	53	63	28	395
Flexural modulus at 23 °C	N/mm^2	3500–4000	4900	62	3700	48,200
at 260–300 °C	N/mm^2	2100–2500	3450	52	1450	
Compressive strength at 23 °C	N/mm^2	200–235	234	138	145	
at 300 °C	N/mm^2	64–68				
Compressive modulus	N/mm^2					1900
Shear strength	N/mm^2	75–80	101			
Impact strength with notch	J/cm	1.7–2.1	2.8	0.86		
Flex fatigue limit (10^7 Hz)	N/mm^2	6.5				
Deformation under load (13.8 N/mm^2 at 60 °C)	%	0.10				
Abrasion (100 h, N_2)	μm	100				
Hardness (Rockwell)		43–44 E	98 E	110 M	99	

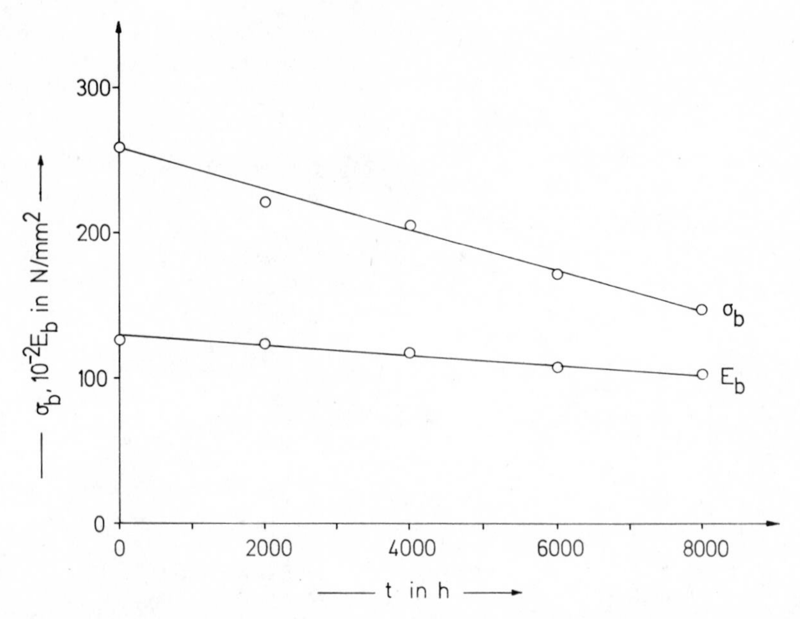

Figure 10-1 Flexural strength σ_b and flexural modulus E_b as a function of time t of glass-fiber reinforced molded sheets made from Kinel 5504 (according to data from Ref. 8).

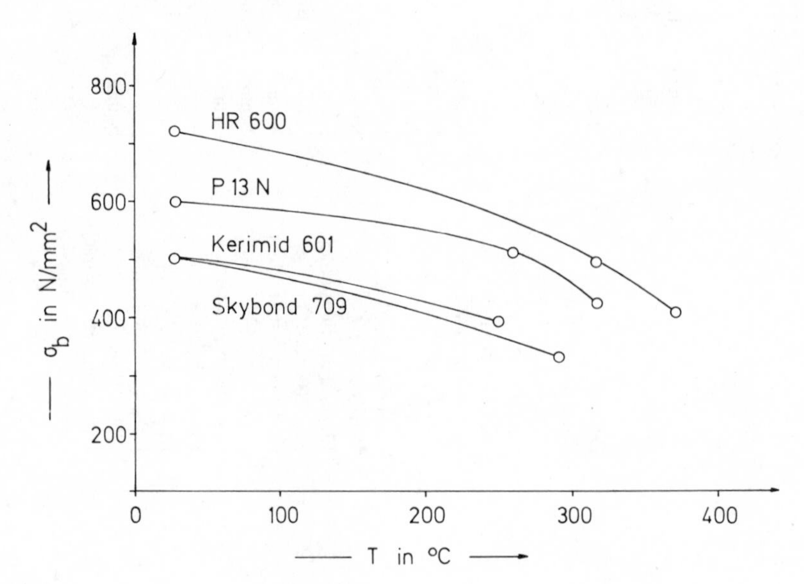

Figure 10-2 Flexural strength σ_b of polyimide/glass fiber mat-laminates as a function of the temperature T (according to Ref. 1).

TABLE 10-5
Properties of polyimide/glass-fiber laminates.

Property	Physical unit	Values measured for Monsanto Skybond 700	Upjohn 2080	Hughes HR 600
Flexural strength				
24 °C	N/mm²	530–600	380	710
0.5 h at 370 °C	N/mm²	320–420		
10 h at 370 °C	N/mm²	140–245		415
Modulus, 24 °C	N/mm²	22,000	27,000	
325 h at 300 °C	N/mm²	22,000		
Tensile strength, 24 °C	N/mm²	400		
335 h at 300 °C	N/mm²	295		
Ultimate elongation, 24 °C	%	1.9		
Impact strength with notch	J/cm		1280	
Water absorption (24 h)	%	0.7		
Dielectric constant				
(10^6 Hz)	1	4.1		
Dielectric strength	kV/cm	71		
Dissipation factor				
(10^6 Hz)	1	0.00445		
Volume resistivity	Ω cm	2.5×10^{15}		
Surface resistivity	Ω	3.4×10^{14}		
Limiting oxygen index				
from top candle	%		90–100	
from bottom	%		70	

300° C for at least 3 h; complete curing requires 6 h at 300° C. Typical properties are listed in Table 10-6. Figure 10-3 shows the decrease in compressive strength with increasing temperature.

Polyimide foams are used as sound insulation at high operating temperatures, such as sound-deadening of jet engines.

10.6 POLYIMIDE FIBERS

The firms Rhône-Poulenc and Upjohn produce polyimide-fibers. Rhône-Poulenc's Kermel®-fibers are polyimideamides, in which R is either a methylene group or an ether group (supposedly the latter). These fibers are yellow-orange. They have a rather poor tensile strength. Post-orientation leads to crystallization and to quite good tensile strength (see Table 10-7). The feel of these fibers is similar to that of acrylic and wool fibers.

The fibers do not melt. They char on heating to about 380–400° C in air or under nitrogen. No harmful gases are given off during burning. The

TABLE 10-6
Properties of Skybond-Foams (according to Ref. 15).

Property	Physical unit	Values measured for type 01	06	12	18
Density	kg/m^3	12	107	218	328
Compressive strength					
5% compression	N/mm^2		0.10	0.66	2.50
10% compression	N/mm^2		0.15	0.95	4.29
25% compression	N/mm^2		0.28	1.65	7.63
Time to a 25% visibility loss during burning	s	35		120	120
Visibility at maximum smoke generation	%	68		97	97
Dielectric constant			1.14	1.30	1.41
Dissipation factor	1		0.0028	0.0028	0.0062

tenacity falls to about one half of the original value at about 260°C, that is the "half tenacity temperature" is somewhat lower than Nomex's 300°C. The fiber shrinks less than 1% when it is heated in boiling water or in dry air

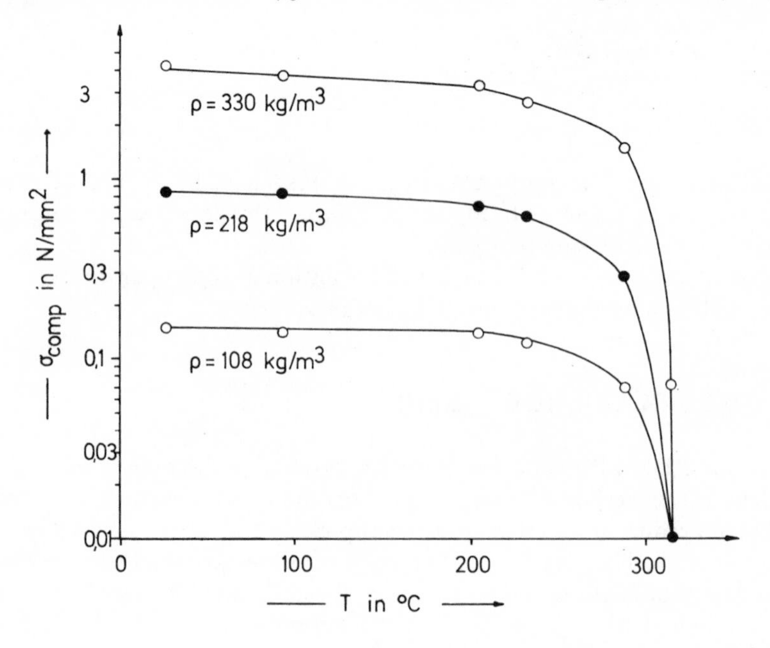

Figure 10.3 Temperature dependence of the compressive strengths σ_{comp} of skybond-foams of varying densities ρ at 10% compression.

TABLE 10-7
Properties of polyimide fibers (type II of Table 10-1) according to data from Ref. 7.

Property	Physical unit	Values measured for R = CH_2	Values measured for R = O
Tensile strength	g/tex	25–50	40–60
ditto after 1000 h at 20 °C	g/tex	7.5–15	32–48
Modulus	g/tex	500–650	600–1000
Ultimate elongation	%	10–25	10–20
Limiting oxygen index, LOI	%	30–32	28–30

(<200° C). It is photo-oxydizable, as are all fibers having an aromatic structure.

Upjohn's fibers are either dry- or wet-spun from solutions of their polyimide 2080. The fiber properties are strongly dependent on the method of spinning and on the stretching ratio (Table 10-8). Wet-spun fibers after 10 min in air at 300° C retain 75 % and at 250° C 60 % of their original tenacities.[16] After 100 h at 300° C they retain 50 %. The limiting oxygen index is 32–36 % depending on the method used.

TABLE 10-8
Properties of Upjohn's polyimide-fibers as a function of spinning method (wet or dry) and stretching ratio (2.5 or 4.0) (according to Ref. 16).

Property	Physical unit	Values measured for wet 2.5	wet 4.0	dry 2.5	dry 4.0
Titer	g/tex	0.35	0.27	0.30	0.14
Tensile strength	N/mm^2	287	341	353	404
Ultimate elongation	%	33	30	44	35
Modulus	N/mm^2	8100	8700	4500	4500
Knot strength	N/mm^2			280	

References

1. N. Bilow, A. L. Landis, R. H. Boschan, R. L. Lawrence and T. J. Aponyi, *ACS Polymer Preprints*, **15**(2), 537 (1974).
2. H. Lee, D. Stoffey and K. Neville, *New Linear Polymers*, McGraw-Hill, New York, 1967, p. 205.

3. N. A. Adrova, J. I. Bessonov, L. A. Laius and A. P. Rudakov, *Polyimides: A new class of heat-resistant polymers*, Israel Program for Scientific Translations, Jerusalem, 1969.
4. M. W. Ranney, Polyimide Manufacture, Noyes Data Corp., Park Ridge, N. J., 1971.
5. Anonym., *Materials Engng.*, —(2), 69 (1974); Anonym., *Kunststoffe*, **64**, 612 (1974).
6. R. Merten, *Angew. Chem.*, **83**, 339 (1971).
7. R. Pigeon and P. Allard, *Angew. Makromol. Chem.*, **40/41**, 139 (1974).
8. Anonym., *Kunststoffe*, **62**, 424 (1972).
9. J.V. Crivello, *J. Polymer Sci.* [*A*-1], **11**, 1185 (1973); *ACS Polymer Preprints*, **14**, 294 (1973).
10. J. V. Crivello, *ACS Polymer Preprints*, **13**, 924 (1972).
11. I. Matsuda, A. Yoshizumi and K. Akiyama, *Kobunshi Ronbunshu* [*Engl. Ed.*], **3**(1), 1148 (1974).
12. H. M. Relles and R. W. Schluenz, *J. Polymer Sci.* [*Chem.*], **11**, 561 (1973).
13. Anonym., *Kunststoffe*, **60**, 401 (1970).
14. Anonym., *Mod. Plastics*, **47**(3), 55 (1970); Anonym., *Kunststoffe*, **61**, 56 (1971).
15. I. Serlin, A. H. Markhart and E. Lavin, *Mod. Plastics*, **47**(7), 120 (1970); Anonym., *Kunststoffe*, **61**, 194 (1971).
16. Product information literature of the firm Upjohn.
17. F. P. Darmory, *ACS Div. Org. Coatings Plast. Chem.*, **34**(1), 181 (1974); M. A. J. Mallet and F. P. Darmory, in R. D. Deanin, ed., *New Industrial Polymers, ACS Symp. Ser.*, **4**, 112 (1974); F. P. Darmory, *ibid.*, **4**, 124 (1974).
18. S. L. Kaplan and S. S. Hirsch, in R. D. Deanin, ed., *New Industrial Polymers, ACS Symp. Ser.*, **4**, 100 (1974).

11 Other Nitrogen-Containing Chains

11.1 AMINO RESINS

11.1.1 Structure

Amino resins (aminoplastics) are condensation products of compounds containing NH-groups with aldehydes or ketones in the presence of nucleophilic compounds (see, for instance, Refs. 1–4).

$$H{-}Y + RR'CO + HN\langle \rightarrow Y{-}C(RR'){-}N\langle + H_2O$$

Commercially urea and melamine are used as the principal NH-compounds and formaldehyde as the carbonyl compound. Nucleophilic compounds are H–, OH–, NH–, SH–, and CH–acids, for instance, hydrogen halides, alcohols and carboxylic acids, urea and malamine, mercaptans and compounds with active methylene groups.

Such condensation products have been used for many years as the resinous components in molding preparations and laminating materials, as binders for lacquers, wood glue, textile finishing and as foams. The use of certain urea/formaldehyde- and melamine/formaldehyde-resins as pigments and fillers is new. Ciba-Geigy makets a urea/formaldehyde resin as a filler and pigment for paper under the tradename Pergopak®.

11.1.2 Synthesis

The melamine/formaldehyde-resins (MF-resins) and the urea/formaldehyde-resins (UF-resins) which are suitable as pigments and fillers are cross-linked products because of the polyfunctionality of the melamine and the urea. They have the base unit

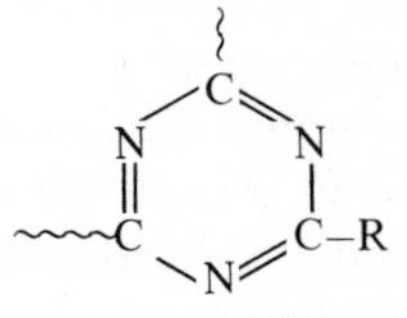

MF: triazine ring (C, N, C, N, C, N) with ~~ substituents and C–R

R = $NHCH_2OH$
$NHCH_2OCH_2NH-$
$NHCH_2NH-$

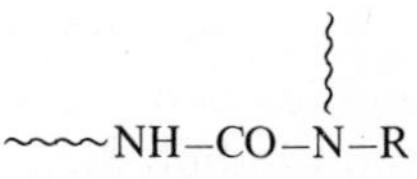

UF: ~~NH–CO–N–R (with ~~ on N)

R = CH_2OH
CH_2NCO-

The UF- and MF-resins to be used as fillers and pigements have to be produced under very specific reaction conditions. The proportion of starting materials and careful control of pH are of utmost importance.[5,6] Typically one starts with 5–30% solution in water. The precipitation of the UF- or MF-resin during the cross-linking polycondensation is controlled by the addition of water-soluble protective colloids. These protective colloids (for instance, Tragacanth, poly(vinyl alcohol)) apparently participate in the polycondensation.[5]

11.1.3 Properties

The MF- and UF-resins precipitated under the above conditions are spherical. Their particle sizes are given as less than 50 nm;[5,6] a particle size of 100 nm is given for Pergopak M.[7] Pergopak M-particles are agglomerized[7] to particles of about 4000–6000 nm (4–6 μm). Bulk densities are about 60 g/1 (Pergopak M) to 100 g/1 (UF-resin as reinforcer for elastomers[8]). The particles are practically ash-free and have a TAPPI degree of whiteness of 97%. They can be stored almost indefinitely in dry surroundings.

The internal surface of these UF- and MF-resins is strongly dependent on the proportions of the starting monomers.[5,6] For instance, UF-resins with starting F/U-ratios of 1.5 are smooth and compact, whereas those with ratios $0.5 < F/U < 1.25$ are rough[6] and have internal surfaces of 55–75 m^2/g.[6,8]

11.1.4 Application

Pergopak M is used as a white pigment and filler for paper.[6,7] It can be added in the form of an aqueous suspension with a solids concentration of less than 10% of the wood pulp practically at any desired step of the paper fabrication.[7] It causes considerably less paper-fabricating machine wear than inorganic fillers. Also the initial wet strength is better. The high degree of whiteness and the good opacity in the dry state make it especially suited for art paper and for the improvement of other papers. Opacity in the wet state is not particularly good. Dimensional stability and receptivity to printing inks are also good.

The use of MF- and UF-resins as reinforcing fillers for the rubber industry has been suggested,[5;6,8] although apparently not yet commercialized.

These white fillers lead to elastomers with high tensile strength at maximum load, high modulus and comparable ultimate elongations as styrene/butadiene-, acrylonitrile/butadiene- and poly(chloroprene)-rubbers filled with other fillers.[8]

UF-resins are also suggested as carriers for herbicides, fungicides and insecticides,[9] because these agents are absorbed by the UF-resins and are only released slowly. UF-resins are decomposed in the ground, and the degradation products are assimilated by plant-life.

References

1. C. P. Vale and W. G. K. Taylor, *Aminoplastics*, Iliffe, London, 1964.
2. A. Bachmann and T. Bertz, *Aminoplaste*, VEB Dtsch. Verlag für Grundstoffindustrie, Leipzig, 1967.
3. R. Vieweg and E. Becker, "Duroplaste", in R. Vieweg and K. Krekeler (Eds.), *Kunststoff-Handbuch*, Vol. 10, Carl Hanser Verlag, Munich, 1968.
4. H. Petersen, Kinetik und Katalyse bei Aminoplastkondensationen, *Chem.-Ztg.*, **95**, 652, 692 (1971).
5. A. Renner, *Makromol. Chem.*, **120**, 68 (1968).
6. A. Renner, *Makromol. Chem.*, **149**, 1 (1971).
7. Pergopak M, Product information literature of the firm Ciba-Geigy.
8. A. Renner, B. B. Boonstra and D. F. Walker, *I.R.I. Conf. on Advances in Polymer Blends and Reinforcement*, Loughborough, Leicestershire, September 15–17, 1969.
9. German Patent 1936748 (January 29, 1970); Ciba-Geigy; inv.: A. Renner and A. Müller; *C. A.*, **72**, 91272r (1970).

11.2 POLY(TEREPHTHALOYLOXAMIDRAZONE)

11.2.1 Structure

Poly(terephthaloyloxamidrazone), PTO, in its metal-chelated form has the base unit

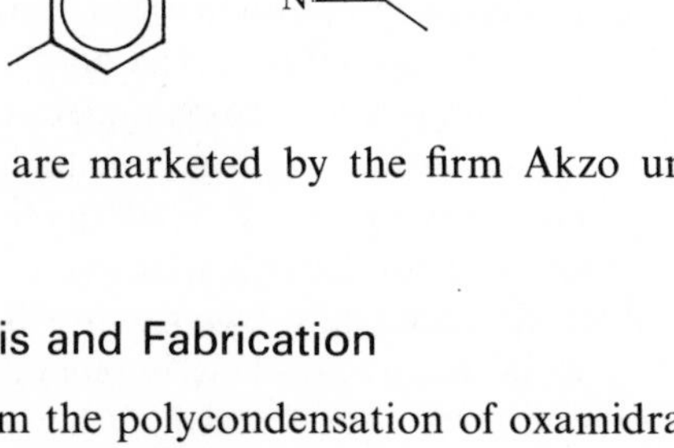

(11-1)

The PTO-chelates are marketed by the firm Akzo under the tradename Enkatherm®.

11.2.2 Synthesis and Fabrication

PTO is formed from the polycondensation of oxamidrazone I with terephthaloyl chloride as shown in Eq. (11-2)

$N{\equiv}C{-}C{\equiv}N$ $\xrightarrow{+2\,H_2N{-}NH_2}$ I

$H_3C{-}C_6H_4{-}CH_3$ $\xrightarrow{+6\,Cl_2,\ -6\,HCl}$ $Cl_3C{-}C_6H_4{-}CCl_3$ $\xrightarrow{+2\,H_2O,\ -4\,HCl}$ $ClOC{-}C_6H_4{-}COCl$ (11-2)

Keto-PTO ⇄ PTO ⇄ Enol-PTO

PTO: $-2\,H_2O$ → Polytriazole; $+2\,OH^{\ominus}$, $-2\,H_2O$ / $+2\,H^{\oplus}$; $-2\,NH_3$ → Polyoxadiazole

↓ chelated PTO

The yellow PTO forms a deep red solution in concentrated base (about 4–6 mol alkali hydroxide/l mol base unit PTO). The alkaline solution can be directly spun into an acid bath, in certain cases under addition of cellulose xanthate. Chelation with the ammonia solution of metal hydroxide is carried out either directly after filament coagulation or on the finished fabric.[1] The filaments are no longer soluble in aqueous alkali after the chelation. Presumably many structures other than the one shown result on chelation. A variety of colors are produced depending on the chelating metal used (Table 11-1); however, white and blue cannot be obtained. Other, darker colors can be printed over these underlying colors.

TABLE 11-1
Composition, color and limiting oxygen index of chelated poly(terephthaloyloxamidrazones).[1-3] Divergent data marked with * are from Ref. 2 and with ** from Ref. 3.

Metal ion				Limiting oxygen index	
Type	Mt/PTO mol/mol	Mt %	Chelate color	conditioned %	dry %
Ca	1	13	brown	50	45
Sr	1/3	10.5	red, orange**	40	34
Ba	1/3		orange		
Sn	1		deep yellow		
Pb	1		deep red		
Bi	1		wine-red		
Cu	2/3 (3/2*)	27.0	olive-green		
Ag	1/2		brown		
Au	1	60	black		
Zn	2 (1*)	20.3	orange	44	37
Cd	2 (1*)	30.2	deep orange		
Zr	1/4		yellow		
Fe	1		black		
Co	3/2	23.5	olive		
Ni	1	18.7	light brown		

11.2.3 Properties and Application

Chelated PTO's have mechanical properties similar to those of rayon (Table 11-2). The thermal properties are, however, outstanding. PTO chelated with

TABLE 11-2
Physical properties of PTO chelated with calcium, strontium or zinc.[2,3]

Property	Physical unit	Values measured for PTO	chelated PTO
Density	g/cm^3		1.7–1.8
Tenacity, conditioned	km	27	17–21
wet	km	15	9–17
Ultimate elongation, conditioned	%	34	30–30
wet	%	36	15–25
Loop tenacity	km	14	9–13
Hot air shrinkage, 3 min, 190°C	%		1–2
Water absorption 40–70% relative humidity	%		9–14
Surface resistivity	Ω		6×10^{12}

zinc, strontium or calcium does not burn at temperatures under 1000°C. It does not shrink or melt.[2] PTO chelated with copper or iron shows an after-glow on combustion, apparently these metals catalyze carbonizing. On burning, no toxic gases are released: only water, carbon dioxide and ammonia.[2,3] PTO chelated with lead is resistant to γ-rays.[1]

Chelated PTO's are recommended for flame-resistant fabrics.

References

1. D. W. van Krevelen, *Angew. Makromol. Chem.*, **22**, 133 (1972).
2. D. Frank, W. Dietrich, J. Behnke, A. Koböck, G. Schuck and M. Wallrabenstein, *Angew. Makromol. Chem.*, **40/41**, 445 (1974).
3. F. C. A. A. van Berkel and H. Grotjahn, *Appl. Polymer Symp.*, **21**, 67 (1973).

11.3 POLY(HYDANTOINS)

11.3.1 Structure and Synthesis

Poly(hydantoins) have the base unit

O X R
–R'–N N–
O

in which R = aliphatic group, R' = aliphatic and/or aromatic group and X = hydrogen or aliphatic group. Poly(hydantoins) have been on the market for some years as electrical insulating varnish[1] and since 1970 as electrical insulating film (from the firm Bayer-Leverkusen).[2]

The poly(hydantoins) used for wire coating I contain both aliphatic and aromatic units in the main chain. They are produced from the reaction of aliphatic diamines with fumaric esters followed by treatment with aromatic diisocyanates.[3] In the following instance $R = C_2H_5$ and $Ar = C_6H_4—CH_2—C_6H_4$ (possibly also $C_6H_4—O—C_6H_4$).

H₂N–R'–NH₂ + 2 CH–COOR ‖ CH–COOR ⟶ ROOC–CH(CH₂–COOR)–NH–R'–NH–CH(CH₂–COOR)–COOR

(11–3)

ROOC–CH(CH₂–COOR)–NH–R'–NH–CH(CH₂–COOR)–COOR + OCN–Ar–NCO ⟶

ROOCCH₂ H O O H CH₂COOR
–R'–N N–Ar–N N–
O O

I

Electrical insulating films are produced from wholly aromatic poly-(hydantoins) II. Of the many possible routes (see Ref. 3) the commercial one apparently uses aromatic diamines, and aromatic diisocyanates in phenol or cresol as solvents. Here Ar is apparently $C_6H_4—O—C_6H_4$.

$$H_2N{-}Ar{-}NH_2 + 2\,ClC(CH_3)_2COOC_2H_5 \xrightarrow[-2\,HCl]{} (C_2H_5OOC{-}C(CH_3)_2{-}NH)_2Ar$$

$$(C_2H_5OOC{-}C(CH_3)_2{-}NH)_2Ar + OCN{-}Ar{-}NCO \longrightarrow$$

$$\longrightarrow -Ar-N<\begin{matrix} C(CH_3)_2{-}COOC_2H_5 \quad C_2H_5OOC{-}C(CH_3)_2 \\ CO{-}NH{-}Ar{-}NH{-}CO \end{matrix}>N- \xrightarrow{-2\,C_2H_5OH}$$

$$\xrightarrow{-2\,C_2H_5OH} \text{[}-Ar-N\overset{H_3C\ CH_3\ O}{\underset{O}{\bigcirc}}N-Ar-N\overset{H_3C\ CH_3\ O}{\underset{O}{\bigcirc}}N-\text{]} \qquad \text{II} \qquad (11\text{–}4)$$

11.3.2 Properties and Application

Highly viscous cresol solutions occur in the preparation of insulating varnish. Commercial varnishes contain, therefore, viscosity-lowering thinners or instead, branching points in the polymer molecule. The varnishes and the films show excellent heat distortion properties (see Table 11-3). Electrical

TABLE 11-3
Properties of poly(hydantoins).

Property	Physical Unit	Values measured for Lacquer I Ref. 1	Film II Ref. 2
Density	g/cm^3		1.27
Heat distortion temperature	°C		270 (tension) 300 (compression)
Continuous service temp.	°C		160
Thermal shock resistance temp.	°C	280	
Tensile strength	N/mm^2		100
Ultimate elongation	%		100
Dielectric constant	1	2.9	
Dissipation factor	1	0.002	
Dielectric strength (40 μm film; 50 Hz; air; ball/plate)	kV/mm	150	200
Behavior at glow discharge	min		2000
Water absorption (24 h)	%		4.5

properties are practically temperature-independent to about 230°C. Stretching of the films noticeably increases their tensile strength at maximum load.

Methylene chloride, dimethylformamide and dimethylacetamide dissolve I; II is resistant to many organic solvents, and to aqueous acids and bases even at their boiling points.[1]

II is compatible with other impregnating resins, for instance with esterimides and unsaturated polyester resins. Poly(hydantoins) are suitable for condenser dielectrics, solder-bath resistant base-material for printed circuits and as insulation for motors having an F heat classification. Films can be coated with copper foil by using thermally stable adhesives.

References

1. W. Dünwald, K. H. Mielke, E. Reese and R. Merten, *Farbe und Lack*, **75**, 1157 (1969).
2. E. Reese, *Kunststoffe*, **62**, 733 (1972).
3. R. Merten, *Angew. Chem.*, **83**, 339 (1971).

11.4 POLY(PARABANIC ACIDS)

11.4.1 Structure

Poly(parabanic acids), also called 2,4,5-triketoimidazolidine polymers, contain the base unit

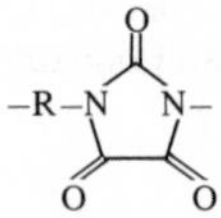

O
–R–N N–
O O

and are, therefore, closely related to the poly(hydantoins). The products developed by Esso have probably $R{=}C_6H_4{-}CH_2{-}C_6H_4$ (PPA–M) and $R{=}C_6H_4{-}O{-}C_6H_4$ (PPA–O); they are not yet commercial. Hoechst produces a number of polymers with varying groups R (see e.g. Ref. 1).

11.4.2 Synthesis

Phenyl isocyanate reacts with hydrogen cyanide in the presence of basic catalysts to give 1,3-phenyl-1-imino-1,3-imidazolidinedione.[2] The reaction can be extended to other monoisocyanates.[3] Dicarbamoyl cyanide can be substituted for hydrogen cyanide.[4] The three-step Esso synthesis leads first to cyanoformamides I, then to poly(*N*-cyanoformylureas) II and finally to poly(iminoimidazolidinediones) III, which are converted to poly(parabanic acids) IV on hydrolysis:[5,6]

$$\sim R{-}NCO + HCN \longrightarrow \underset{\text{I}}{\sim R{-}NH{-}C(=O){-}CN} \xrightarrow{+\,OCN\sim} \underset{\text{II}}{\sim R{-}N(C(=O){-}CN){-}C(=O){-}NH\sim} \qquad (11\text{-}5)$$

$$\sim R{-}N(C(=O){-}CN){-}C(=O){-}NH\sim \longrightarrow \sim R{-}N\underset{\text{ring: }C(=O),\ C(=NH)}{\overset{C=O}{\frown}}N\sim \xrightarrow[-\,NH_3]{+\,H_2O} \sim R{-}N\underset{\text{ring: }C(=O),\ C(=O)}{\overset{C=O}{\frown}}N\sim$$

Soluble polymers are formed in polycondensations with tertiary amines or pyridine if hexamethylene diisocyanate is the starting material. All other diisocyanates give cross-linked products. With sodium cyanide as the catalyst soluble polymers with high molecular weights are formed, even with these other diisocyanates.[7]

The Hoechst synthesis starts with oxamide esters and isocyanates or capped isocyanates[8]

$$\sim NH{-}CO{-}CO{-}OR + 2 \sim B{-}NCO \rightarrow \sim A{-}N\underset{C=O}{\overset{O=C{-}C=O}{\frown}}N{-}B\sim + \sim B{-}NH{-}CO{-}OR \qquad (11\text{–}6)$$

11.4.3 Properties, Fabrication and Application

The Esso products are noncrystalline polymers with relatively high glass temperatures of 200–300°C.[7,9] According to thermogravimetric measurements they are stable to about 400°C (PPA–M) or 470°C (PPA–E). PPA–E shows somewhat greater oxidation resistance than PPA–M.[9]

Both PPA–M and PPA–E are soluble in dimethylformamide, *N*-methylpyrrolidone and pyridine. PPA–M is also soluble in cyclic ketones and in lactones. Hydrocarbons, halogenated hydrocarbons, ethers and acids do not attack poly(parabanic acids), and aliphatic ketones do so only slightly. Bases can cause stress cracking.[9]

The stress/strain curve shows the typical yield of tough plastics even at −46°C. Even at 260°C poly(parabanic acids) still have the strength shown by poly(tetrafluoroethylene) at room temperature.[9] They have very good creep resistance: only a 1% increase in length is measured after 400 h at 100°C and under a load of approximately 21 N/mm^2. Poly(parabanic acids) can be stretched at elevated temperatures to clear, highly oriented films. The electrical properties are comparable to those of polyimides.

TABLE 11-4
Properties of poly(parabanic acids) at 23°C (according to Ref. 9, compression molding fabrication).

Property	Physical unit	Values measured for PPA-M	Values measured for PPA-E
Density	g/cm^3	1.30	1.38
Glass temperature	°C	290	299
Heat distortion temperature	°C	270	282
Coefficient of linear thermal expansion	K^{-1}	5.7×10^{-5}	3.3×10^{-5}
Tensile yield strength	N/mm^2	99	106
Yield	%	8–10	8–9
Tensile strength	N/mm^2	99	
Ultimate elongation	%	8–50	15–60
Modulus	N/mm^2	2500	2700
Flexural modulus	N/mm^2	2820	
Flexural strength	N/mm^2	127	
Impact strength with notch	J/m	55–75	
Rockwell hardness		M 95	
Water absorption (24 h)	%	1.8	
Dielectric constant	1	3.82	3.60
Volume resistivity	Ω cm	$>10^{17}$	$>10^{17}$
Dissipation factor	1	0.0040	0.0027
Surface resistivity	Ω	$>10^{16}$	$>10^{16}$
Arc resistance	s	125	125
Dielectric strength	kV/cm	2400	2400

The Esso products can be fabricated by film-casting and compression molding into films, coatings and electrical insulating materials. Solutions are also usable as adhesive materials.

The Hoechst products are normally produced and sold as solutions in N-methylpyrrolidone with 15–55% solid content. The finished products show high thermal stability (weight loss smaller than 6% at 400°C) and high softening temperatures of 390°C and higher. The surface hardness is about eight times higher than that of comparable polyamideimide polymers. Polymers with inherent viscosities of 50 cm^3/g are used for wire lacquers, those with inherent viscosities of 100 cm^3/g for films, and those with inherent viscosities of 85–170 cm^3/g for fibers.[1]

References

1. H. Brandrup, Hoechst AG, personal communication, April 15, 1976.
2. W. Dieckmann and H. Kämmerer, *Chem. Ber.*, **38**, 2977 (1905).

3. T. L. Patton, *J. Org. Chem.*, **32**, 383 (1967).
4. A. Oku, M. Okana and R. Oda, *Makromol. Chem.*, **78**, 186 (1964).
5. U.S. Patent 3547897 (February 3, 1969); Esso Res. Eng.; inv.: T. L. Patton; *C.A.*, **73**, 99411b (1970)
6. U.S. Patent 3591562; Germ. Pat. 2003938 (August 13, 1970); Esso Res. Eng.; inv.: T. L. Patton; *C. A.*, **73**, 99411 (1970).
7. T. L. Patton, *ACS Polymer Preprints*, **12**, 162 (1971).
8. Austrian Patent 300792 (June 18, 1971/August 10, 1972); Reichhold-Albert-Chemie, Hamburg (a subsidiary of Hoechst AG).
9. D. J. Henderson, B. H. Johnson and T. L. Patton, *SPE 30th Ann. Tech. Papers*, **18**, pt. 2, 669 (1972).

11.5 HYDROPHILIC POLYURETHANE-FOAMS

In the usual production of polyurethane-foams three components (isocyanate, polyol, water) are mixed together in precise quantities. W. R. Grace and Company has now developed a hydrophilic polyisocyanate which gives a foam directly on addition of water, and the water need not be measured exactly (about 35–200 parts water per 100 parts polyisocyanate). The new product is marketed under the tradename Hypol®.[1] Hypol 3000 is an all-purpose product which is suitable for the manufacture of both soft and rigid foams. Hypol 2000 is a specialty product for the manufacture of very soft foam (Angel Foam®).

Foaming need not be catalyzed. The foams can have densities between 32 and 320 kg/m^3.[2] Curiously, both the rigid and the soft foams are open-celled. Other commercial rigid foams normally have closed-cells. The foams have LOI's of 25–35 and are, therefore, moderately flame retardant. Additional flame retardant leads to LOI's of up to about 70, and at the same time reduces smoke on burning. The foams char and shrink on application of flame.

Hypol-foams are hydrophilic and can absorb appreciable amounts of water. Their mechanical properties are supposed to be comparable to those of conventional polyurethane-foams.

References

1. Anonym., Formed Fabrics Industry, October 1974.
2. Product information literature of the firm W. R. Grace and Co.

11.6 TRIAZINE POLYMERS

Ciba-Geigy Corp. has developed a new class of thermosetting *N*-cyanosulfonamide polymers, designated as NCNS polymers.[1,2] Reaction of primary and secondary biscyanamides leads to a soluble prepolymer (I). Curing at 150°C leads to formation of cross-links composed of *s*-triazine rings (II). The mechanism of the reactions is somewhat uncertain but is tentatively given as

$$\mathrm{NC{-}NR{-}A{-}NR{-}CN + NC{-}NH{-}A'{-}NH{-}CN} \rightarrow$$

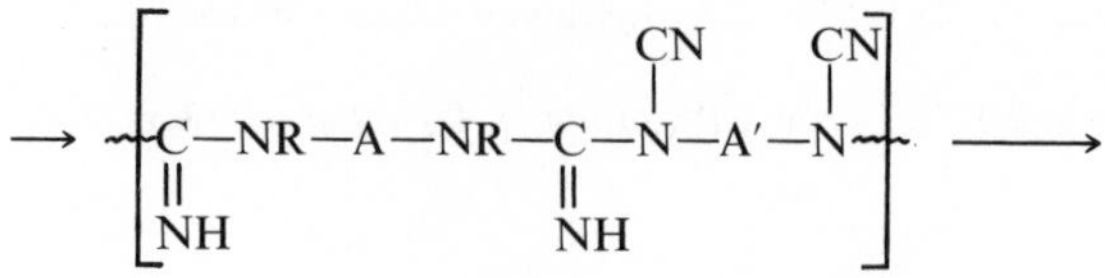

$$\rightarrow \left[\underset{\underset{\textstyle\mathrm{NH}}{\|}}{\mathrm{C}}{-}\mathrm{NR{-}A{-}NR}{-}\underset{\underset{\textstyle\mathrm{NH}}{\|}}{\mathrm{C}}{-}\overset{\overset{\textstyle\mathrm{CN}}{|}}{\mathrm{N}}{-}\mathrm{A'}{-}\overset{\overset{\textstyle\mathrm{CN}}{|}}{\mathrm{N}} \right] \longrightarrow$$

I

$$\left[\mathrm{NR{-}A{-}}\underset{\underset{\textstyle\mathrm{R}}{|}}{\mathrm{N}}{-}(\text{s-triazine ring: N, N, N}){-}\mathrm{NH{-}A'{-}NH} \right]$$

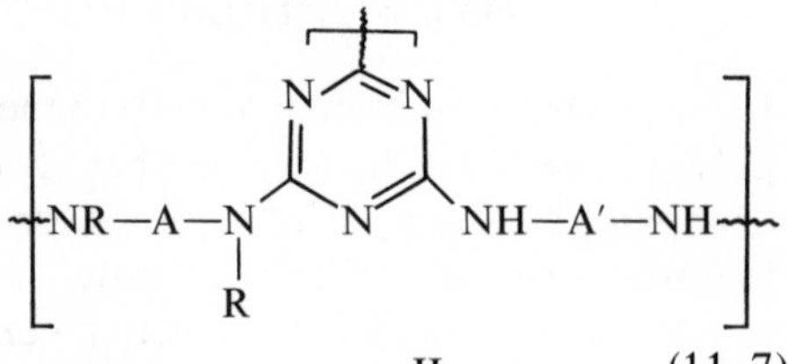

II (11–7)

where A and A′ are aromatic groups and R are electrophilic groups which increase the resin processability such as pendant aryl sulfonyl groups.[2] The same group reported previously on *s*-triazine based poly(benzimidazoles) and poly(imides).[3]

NCNS resins are prepared from 1:1 and 1:2 molar ratios of secondary to primary biscyanamides in low molecular weight alcohols or ketones. The resulting A-stage prepolymers have to be B-staged for about 30 min at 100–120°C to increase their melt viscosity.

The unfilled NCNS resins can be compression molded to hard, transparent, rigid articles of low elongation (Table 11-5). They are compatible with up to 50% chopped glass fibers and particulate silica. The 1:1 resin may be utilized as an injection or transfer molding resin, the 1:2 resin as a rapid-cycle compression molding resin.

NCNS resins have wide applicabilities as laminates with glass fabrics, graphite fibers, and Kevlar fabrics (Table 11-5) for low-void composites, circuit boards or honeycomb structures.[4]

References

1. Anonym. *Chem. Tech.*, **5**, 511 (1975).
2. R. J. Kray, *Mod. Plastics*, **52**(4), 72 (1975).
3. R. J. Kray, R. Seltzer and R. A. E. Winter, *ACS Coatings and Plastics Preprints*, **31**(1), 569 (1971).
4. R. J. Kray, *Soc. Plast. Ind., 30th Ann. Conf.* (1975), Sect. 19-C, 1.

TABLE 11-5
Properties of NCNS resins.[2]

Property	Physical units	Property values					
		Compression moldings		Laminates			
				glass fabric		graphite fiber	Kevlar fabric
		1:1	1:2	1:1	1:2	1:2	1:2
Resin content	%	100	100	34.4	32.5	40.1	47.5
Molding conditions (21.5 N/mm^2)	°C	204	149				
Postcure (232 or 260°C)	h	5	2				
Density	g/cm^3	1.40	1.31	1.90	1.90		1.40
Glass transition temperature (after cure at 232°C)	°C	280	300				
Flexural strength (23°C)	N/mm^2	82	117	570	620	1400	370
(232°C)	N/mm^2			490	525	940	
Flexural modulus (23°C)	N/mm^2	5730	4420	23,700	26,700	102,500	25,900
(232°C)	N/mm^2			19,900	23,400	96,600	
Tensile strength	N/mm^2		49				
Tensile modulus	N/mm^2		4100				
Elongation at break	%		1.4				
Short beam strength (span to depth ratio 4:1)							
(23°C)	N/mm^2			48.1	53.0	88.9	
(232°C)	N/mm^2			45.3	43.7		
Barcol hardness	—	56	51				
Water absorption (24 h at 25°C)	%	0.49	0.64				
Limiting oxygen index	%		29				

12 Inorganic Polymers

12.1 POLY(PHOSPHAZENES)

12.1.1 Structure

Poly(phosphazenes) are polymers having alternating phosphorus- and nitrogen-atoms in the main chain

$$\mathrm{+P(XX')=N+}$$

The fully inorganic poly(phosphazenes) with $X = X' = Cl$ have been known since 1895.[1] They are commonly named poly(phosphonitrilic chloride). They decompose rapidly in moist air; and consequently, they achieved no commercial importance. Substitution of chlorine atoms with fluoroalkoxy groups[2] opened the way to commercially useful products. These fluorine-containing poly(phosphazenes) were produced in experimental quantities by Firestone Tire and Rubber Co. under the tradename PNF (=phosphonitrilic fluoroelastomers). For a review, see Refs. 3 and 19.

12.1.2 Synthesis

Hexachlorocyclotriphosphazene $(PCl_2N)_3$ and octachlorocyclotetraphosphazene $(PCl_2N)_4$ "thermally" polymerize at elevated temperatures to high molecular weight products, for instance,

$$n/3\ (\mathrm{PCl_2N})_3\ \text{(cyclic: } \mathrm{Cl_2P}, \mathrm{PCl_2}, \mathrm{PCl_2} \text{ ring with N)} \rightleftharpoons \mathrm{+P(Cl_2)=N+_n} \qquad (12\text{–}1)$$

This thermal polymerization is not reproducible. Cross-linked product results under certain conditions at conversions greater than 50% and with near certainty at conversions greater than 75%,[4] to give the so-called inorganic rubber. Cross-linking is unlikely at conversions under 35%. The poor reproducibility presumably results from the fact that the polymerization is not spontaneously thermal, but rather is initiated cationically by impurities which are difficult to remove.[4] The polymerization equilibrium shown in Eq. (12-1) is not applicable to the synthesis of poly(organophosphazenes). Under the conditions necessary for thermal activation the polymerization equilibrium is on the side of the monomers, that is of the cyclic trimers and tetramers.[2] On the other hand the catalyzed polymerization of hexaphenylcyclotriphosphazene to high molecular weight poly(hexaphenylphosphazenes) is claimed.[5] Poly(phosphazenes) with organic side groups are prepared in polymer-analogous reactions on poly(dichlorophosphazene) through chlorine displacement.[2]

$$\text{+P(Cl)}_2\text{N+}_n + 2\,n\,\text{NaOCH}_2\text{CF}_3 \xrightarrow[\text{Benzene}]{\text{THF}} \text{+P(OCH}_2\text{CF}_3)_2\text{N+}_n + 2\,n\,\text{NaCl} \qquad (12\text{–}2)$$

The homopolymers are crystalline (see below) and the copolymers are amorphous. The commercially available poly(fluoroalkoxyphosphazenes) are synthesized from solutions of uncross-linked, soluble, high molecular weight poly(dichlorophosphazenes) in benzene by treatment with a mixture of sodium fluoroalkoxide in tetrahydrofuran.[4,6–8] The resulting products are probably statistical copolymers, since ligand exchange is quite fast even under mild reaction conditions.[4,9]

12.1.3 Fabrication

The commercial PNF-polymers with about 65% OCH_2CF_3— and 35% $OCH_2(CF_2)_4H$— groups and a very small amount of an unspecified reactive group have a loss modulus at 100°C similar to that of butadiene/styrene-rubber or EPDM. At higher shear gradients it is more like that of silicone rubber[4,8,17] (Figure 12-1). As a result, the PNF-polymers have good green strength.[8] They feel tough and show no tendency to cold flow. Storage modulus is similar to that of SBR. Therefore, the PNF-polymers can be fabricated in the same way as other elastomers.

PNF can be vulcanized with organic peroxides, sulfur and accelerators or through high energy irradiation.[8] Elongation and compression set decrease with increasing amount of peroxide, that is with increasing cross-linking (Table 12-1). At the same time the modulus and the hardness increase. Tensile strength at maximum load is higher than that of silicones and reaches that of conventional elastomers. Vulcanization with sulfur/accelerator-systems also leads to products with outstanding mechanical strength; however, these products have poorer properties with respect to compression set and heat aging than do peroxide cross-linked materials.[8]

The poly(fluoroalkoxyphosphazenes) can be supplied with conventional reinforcing fillers such as silica, carbon-black or other inorganic materials.

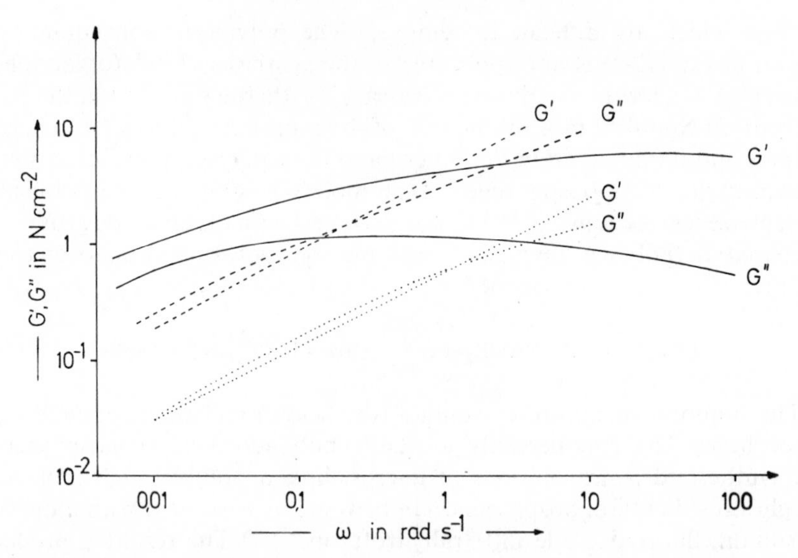

Figure 12-1 Storage modulus G' and loss modulus G'' of PNF (——) as a function of shear rate ω at 100 °C. For comparison, the corresponding values for an EMDM-rubber (——) and a silicone-rubber (Silastic 430) (· · ·) (according to data from Ref. 8) are given.

Increasing quantities of these fillers lead to increasing modulus and compression set while the elongation remains practically constant (see Table 12-1).

12.1.4 Properties

The homopolymeric poly(organophosphazenes) are crystalline materials[2,10–14] with relatively high melting temperatures (Table 12-2). The copolymers are amorphous.[6,7,13] Many poly(organophosphazenes) have transition temperatures[13] other than glass- and melt-temperatures (see Table 12-2).

Poly(organophosphazenes) have molecular weights of several million. The molecular weight distribution is broad and has a long, low molecular weight tail.[2,4,8,14,15] These polymers, therefore, are not present in thermodynamic equilibrium, otherwise exchange reactions would give a Schulz–Flory distribution.

The uncross-linked poly(dichlorophosphazenes) are soluble in benzene, tetrahydrofuran, carbon tetrachloride and ethylene glycol dimethyl ether (diglyme).[8] The poly(1,1,1-trifluoroethoxyphosphazenes) are soluble in acetone, butanone, ethyl acetate and diglyme, but not in many other organic solvents.[2] A poly(organophosphazene) with 65 mol-% OCH_2CF_3 and 35 mol-% $OCH_2(CF_2)_4H$ is soluble in dimethylformamide and methyl isobutyl ketone,[4,8] and a copolymer with OCH_2CF_3 and $OCH_2C_3F_7$ (1:1) is soluble in trichlorotrifluoroethane.[6]

TABLE 12-1
Properties of filled and dicumyl peroxide cross-linked poly(fluoroalkoxyphosphazenes) with 0.65 mol CF_3CH_2O/0.35 mol $H(CF_2)_4CH_2O$. Vulcanization 30 min at 160°C, post-vulcanization 24h at 100°C. Silanox = organosilyl silicon dioxide, Stam Mag ELC = magnesium oxide, Dicup 40 C = dicumyl peroxide.

	Physical unit	Composition					
PNF		100	100	100	100	100	100
Silanox 101			20	30	40	30	30
MT Black		25					
Stam Mag ELC		6	6	6	6	6	6
Dicup 40 C		2	1	1	1	3	6
Modulus at 100% elongation	N/mm^2	4.2	2.0 ± 0.1	3.2	4.4	7.7	10.0
Tensile strength	N/mm^2	7.1	10.2 ± 1.0	14.1	16.7	11.3	10.9
Ultimate elongation	%	125	185 ± 15	170	175	117	105
Shore hardness, A			30	60	76	70	72
Compression set (70 h at 135°C)	%		36	56	80	34	33
Low temperature torsion (ASTM D-1053)							
T_5	°C	−52	−38				
T_{10}	°C	−55	−42				
T_{100}	°C	−63	−53				
Glass transition temperature	°C	−64	−43				

Cross-linked poly(fluoroalkoxyphosphazenes) swell only slightly in hydrocarbons,[8] for example 3.5 vol.-% in ASTM-liquid A and 12.8 vol.-% in ASTM-liquid C after 100 h at 23°C. They are very stable to hydrolysis.[8]

According to thermogravimetric measurements,[14,16] the decreasing order in the thermal stability of uncross-linked poly(alkoxyphosphazenes) is

$$OCH_2CF_3 > OC_6H_5 > OC_2H_5 > OCH_3$$

The tensile strength at fracture of cross-linked, commercial PNF is only slightly affected after 100 h at 150°C.[8] However, it falls to about half after the same time period at 205°C. The modulus is only slightly altered in the same temperature range. These data lead to the conclusion that commercial PNF can be used for up to 1000 h at 150°C. The cause for the drop in tensile strength at elevated temperatures is not exactly known. It is neither the result of oxidation nor of polymer/monomer equilibrium, but probably rather of exchange equilibria. The cross-linked PNF's have excellent low-temperature strength (see Table 12-2). The reinforcing fillers, carbon black

TABLE 12-2

Melting points T_M and statistical glass temperatures T_G of poly(phosphazenes) $\text{-(-}P(RR')N\text{-)-}_n$ having varying substituents R and R' or R = R'.

R or R'	T_M °C	T_G °C
Cl	−9 [13]	−67 [13]; −63 [16]
OCH_3		−77 [2]; −76 [16]
OC_2H_5		−94 [2]; −84 [16]
OCH_2CF_3	240 [2]; 238 [13]	−66 [2, 16]; −70 [14] −53 [13]
$OCH_2CF_3/OCH_2C_3F_7$ (1:1)	amorphous [13]	−77 [6]; −64 [13]
$OCH_2C_3F_7$	> 250 [13]	
$OCH_2C_7F_{15}$		−40 [14]
OC_6H_5		−8 [2; 16]
$OC_6H_4F(p)$		−14 [14]
$OC_6H_4CF_3(m)$	330 [14]	−35 [14]
$OC_6H_4Cl(p)$	> 350 [14]	−12 [14]
$OC_6H_4Cl_2(2,4)$	210 [14]	2 [14]
$OC_6H_4C_6H_5(p)$	> 350 [14]	43 [14]

and silica, do not strongly affect the brittleness temperature, rather they stiffen the material and give it a higher compression set (see Table 12-1).

Poly(aryloxyphosphazenes)[12,18] and poly(fluoroalkoxyphosphazenes)[8] show flame resistant behavior. Silanox reinforced PNF has a LOI limit of 50, and surprisingly, carbon-black reinforced material one of 64. There are no data about smoke generation and the products of combustion.

12.1.5 Application

Poly(fluoroalkoxyphosphazenes) have been developed for applications requiring good solvent resistance, good low temperature flexibility and good mechanical strength over a broad temperature range, from about −60 and +200°C. Typical applications are as O-rings, gaskets, fuel hoses and vibration damping services. Here they compete with fluorosilicones, which have poorer dynamic properties.

References

1. H. N. Stokes, *Amer. Chem. J.*, **17**, 275 (1895); **19**, 782 (1897).
2. R. Allcock and R. L. Kugel, *J. Amer. Chem. Soc.*, **87**, 4216 (1965).
3. H. R. Allcock, *Phosphorus-Nitrogen Compounds*, Academic Press, New York and London, 1972.
4. D. P. Tate, *J. Polymer Sci.* [*Symp.*], **48**, 33 (1974).

5. V. V. Korshak, V. V. Kireev, M. A. Eryan and I. B. Telkova, USSR-Pat. 417452 (February 28, 1974); *C. A.*, **81**, 170234u (1974).
6. S. H. Rose, *J. Polymer Sci. B* [*Letters*], **6**, 837 (1968).
7. U.S. Patent 3702833 (November 14, 1972); Horizons Inc.; inv.: S. H. Rose and K. A. Reynard; *C. A.*, **78**, 98764k (1973).
8. G. S. Kyker and T. A. Antkowiak, *Rubber Chem. Tech.*, **47**, 32 (1974).
9. H. R. Allcock and G. Y. Moore, *Macromolecules*, **5**, 231 (1972).
10. M. Kajiwara, H. Saito and T. Saito, *Kobunshi Kagaku*, **30**, 374 (1973).
11. R. E. Singler, G. L. Hagnauer, N. S. Schneider, B. R. Laliberte, R. E. Sacher and R. W. Matton, *J. Polymer Sci.* [*Chem.*], **12**, 433 (1974).
12. Anonym., *Kunststoff-Berater*, **19**, 208 (1974).
13. T. M. Connelly, Jr. and J. K. Gillham, *ACS Polymer Preprints*, **15**(2), 458 (1974).
14. G. Allen, C. J. Lewis and S. M. Todd, *Polymer* [*London*], **11**, 44 (1970).
15. G. L. Hagnauer and N. S. Schneider, *J. Polymer Sci.* [*A-2*], **10**, 699 (1972).
16. H. R. Allcock, R. Kugel and K. J. Valan, *Inorg. Chem.*, **5**, 1709 (1966).
17. C. W. Macosko and F. C. Weissert, ASTM-Meeting, Philadelphia, June 6, 1973; cited in Ref. 8.
18. R. E. Proodian, *SPE Tech. Papers*, **19**, 714 (1973).
19. R. E. Singler, N. S. Schneider and G. L. Hagnauer, *Polymer Engng. Sci.*, **15**, 321 (1975).

12.2 POLY(CARBORANE SILOXANES)

12.2.1 Structure

Poly(carborane siloxanes) contain both *m*-carborane groups and siloxane groups in the main chain

$$-SiR_2-CB_{10}H_{10}C-(SiR_2-O-)_x- \qquad I$$

They were first described in 1966.[1] Various types have been brought onto the market since 1971 under the tradename Dexsil® by the Olin Corp.[2] Union Carbide is also active in this area,[3] but as yet appears to have no products on the market.

The various Dexsils are designated by three figure numbers. The first number gives the number of siloxane groups in the structural element. The last number is a code for the substituents: A 0 means $R=CH_3$; a number other than zero means other substituents. Dexsil 300 is thus a poly(carborane siloxane) with methyl substituents and $x=3$ in I. Dexsil 202 has $x=2$ siloxane groups and phenyl groups in addition to the methyl groups.

12.2.2 Synthesis

Usually decaborane $B_{10}H_{14}$ (II) is the starting material for the polymer synthesis; however, pentaborane B_5H_9 can also be used. Acetylene is added to decaborane, and the resulting *o*-carborane (=1,2-dicarbaclorclovododecaborane) (III) rearranges at 475°C to *m*-carborane (IV). On treatment with butyl-lithium *m*-carborane gives the *m*-dilithium compound $LiCB_{10}H_{10}CLi$ (V):

$$\text{II} \xrightarrow{+C_2H_2} \text{III} \longrightarrow \text{IV} \xrightarrow{\text{BuLi}} \text{V} \qquad (12\text{-}3)$$

The *m*-dilithium compound (V) is used for a series of syntheses. Reaction with dichlorosilanes R_2SiCl_2 leads to a carborane with two chlorosilane endgroups (VI), which then can be converted with, for instance, methanol into the corresponding dimethoxy compound (VII). Polycondensation of (VI) with (VII) in the presence of ferric chloride gives thermoplastic polymers with glass temperatures of 77°C and melting temperatures of 464°C:

$$n\ Cl{-}\underset{CH_3}{\overset{CH_3}{Si}}{-}CB_{10}H_{10}C{-}\underset{CH_3}{\overset{CH_3}{Si}}{-}Cl \ + \ n\ CH_3O{-}\underset{CH_3}{\overset{CH_3}{Si}}{-}CB_{10}H_{10}C{-}\underset{CH_3}{\overset{CH_3}{Si}}{-}OCH_3 \longrightarrow \qquad (12\text{-}4)$$

(VI) (VII)

$$\xrightarrow[-\,2\,n\,CH_3Cl]{FeCl_3} \left[\underset{CH_3}{\overset{CH_3}{Si}}{-}CB_{10}H_{10}C{-}\underset{CH_3}{\overset{CH_3}{Si}}{-}O \right]_n$$

(VIII)

The polymers are lightly cross-linked as a result of the high polycondensation temperatures,[2,3] which can lead to fabrication difficulties. Elastomers are obtained when the silyl groups in (VI) and (VII) are replaced with siloxane groups. In contrast to the carborane compounds with chlorosilane groups (VI), the carborane compounds with siloxane groups undergo a polycondensation with water, even at 2°C:[2]

$$LiCB_{10}H_{10}CLi \ + \ 2\ Cl{-}SiR_2{-}O{-}SiR_2{-}Cl \longrightarrow$$

$$\longrightarrow Cl{-}SiR_2{-}O{-}SiR_2{-}CB_{10}H_{10}C{-}SiR_2{-}O{-}SiR_2{-}Cl \ + \ 2\ LiCl \qquad (12\text{-}5)$$

$$Cl{-}SiR_2{-}O{-}SiR_2{-}CB_{10}H_{10}C{-}SiR_2{-}O{-}SiR_2{-}Cl \ + \ 2\ H_2O \longrightarrow$$

$$\longrightarrow \lbrace SiR_2{-}CB_{10}H_{10}C{-}({-}SiR_2{-}O{-})_3 \rbrace \ + \ 2\ HCl$$

(IX) (12–6)

The structure of these poly(carborane siloxanes) can be varied considerably by incorporating phenyl- or vinyl-group containing compounds into the polymer. Olin Corp. produces, for example, a compound with the following composition:[4]

$$\left(-\underset{CH_3}{\overset{CH_3}{|}}\!\!\!\!\!Si-CB_{10}H_{10}C-\underset{CH_3}{\overset{CH_3}{|}}\!\!\!\!\!Si-O-\right)_{50}\left(-\underset{C_6H_5}{\overset{C_6H_5}{|}}\!\!\!\!\!Si-O-\right)\left(-\underset{CB_{10}H_{10}C-CH=CH_2}{\overset{CH_3}{|}}\!\!\!\!\!Si-O-\right)_1 \quad (X)$$

The polycondensation of substances with silanol endgroups does not yield high enough molecular weights, even when sulfuric acid is used as the catalyst. Bases, even very weak ones such as calcium oxide, split the silicon/carbon bond. According to work by Union Carbide[3] high molecular weight polymers can be produced by the condensation of the hydroxyl groups of bis(hydroxydimethylsilyl)-*m*-carborane with bis(*N*-pyrrolidino-*N*′-phenylureidodialkylsilanes:

$$HO-\underset{CH_3}{\overset{CH_3}{|}}\!\!\!\!\!Si-CB_{10}H_{10}C-\underset{CH_3}{\overset{CH_3}{|}}\!\!\!\!\!Si-OH + \left(C_4H_8N-CO-\underset{C_6H_5}{|}\!\!\!\!N-\right)_2 SiR'R'' \longrightarrow \quad (12\text{–}7)$$

$$\longrightarrow \left[-\underset{R''}{\overset{R'}{|}}\!\!\!\!Si-O-\underset{CH_3}{\overset{CH_3}{|}}\!\!\!\!\!SI-CB_{10}H_{10}C-\underset{CH_3}{\overset{CH_3}{|}}\!\!\!\!\!Si-O-\right] + 2\ C_4H_8N-CO-NH-C_6H_5$$

(XI)

where either $R' = R'' = C_6H_5$ or $R' = C_6H_5$ and $R'' = CH_3$. No cross-linked product is formed, in contrast to reaction (12-4). Incorporation of small quantities of methyl vinyl siloxane affords vulcanizable products. A typical product (XII) has the following composition:

$$\left(-\underset{CH_3}{\overset{CH_3}{|}}\!\!\!\!\!Si-CB_{10}H_{10}C-\underset{CH_3}{\overset{CH_3}{|}}\!\!\!\!\!Si-O-\underset{CH_3}{\overset{CH_3}{|}}\!\!\!\!\!Si-O-\right)_{66}\left(-\underset{CH_3}{\overset{CH_3}{|}}\!\!\!\!\!Si-CB_{10}H_{10}C-\underset{CH_3}{\overset{CH_3}{|}}\!\!\!\!\!Si-O-\underset{C_6H_5}{\overset{CH_3}{|}}\!\!\!\!\!Si-O-\right)_{33}\left(-\underset{CH_3}{\overset{CH_3}{|}}\!\!\!\!\!Si-CB_{10}H_{10}C-\underset{CH_3}{\overset{CH_3}{|}}\!\!\!\!\!Si-O-\underset{CH=CH_2}{\overset{CH_3}{|}}\!\!\!\!\!Si-\right)_1 \quad (XII)$$

12.2.3 Properties and Application

Poly(carboranesiloxanes) with two or more siloxanyl groups (x = 2 in I) are elastomers. The glass temperatures of a large number of compounds have been extensively investigated (see Refs. 5 and 6).

The polymer marketed as an elastomer by Olin has only half the ultimate elongation at room temperature of a similarly filled natural rubber. The ultimate elongation exceeds that of natural rubber only above 150° C, where it remains more or less constant. Therefore, this poly(carboranesiloxane) can be used at operating temperatures up to 250° C. At these temperatures antioxidants such as ferric oxide must be added. Short term use up to even 400° C appears possible.[4] Polymer (XII) has similar aging resistance (see figure 12-2).

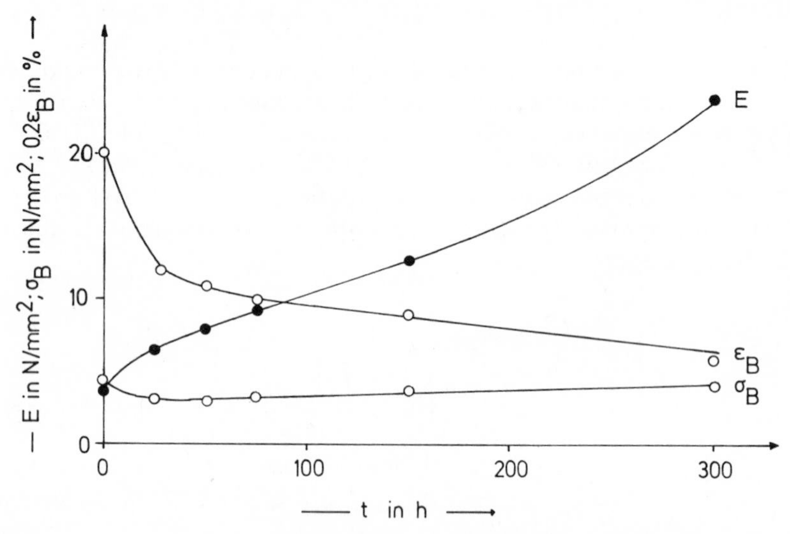

Figure 12-2 Tensile strength σ_B, ultimate elongation ε_B, and modulus E, each measured at 25°C, as a function of aging time t at 315°C for poly(carboranesiloxane) (XII). The elastomer (100 parts) was filled with 30 parts pyrogenic silicic acid, cross-linked with 2.5 parts dicumyl peroxide and contained 2.5 parts ferric oxide (according to data from Ref. 3).

Table 12-3 lists a few mechanical properties. Figure 12-3 shows the temperature dependence of these values for polymer (XII). Polymer (X) filled with precipitated silicic acid has a thermal stability which corresponds to an ASTM-heat class K or L, and with very pure pyrogenic silicic acid even class R or S.[4]

The principal advantage of these new elastomers is found in their excellent long-term thermal stability. For a few, heat distortion temperatures of up to even 500° C have been reported.[2] Thermal stability increases on the substitution of phenyl groups into the polymer, as is already known with the poly(siloxanes).

Potential applications are, therefore, as gaskets and O-rings for motors and exhaust pipes, wire insulation, coated glass fabric for hot air regulation in jet engines and in other areas for elastomers at high temperatures. Solu-

TABLE 12-3
Properties of some poly(carborane siloxanes) filled with silicic acid and vulcanized.

Property	Physical unit	Values measured for X	Values measured for XII
Tensile strength	N/mm^2	10	4.3
Ultimate elongation	%	260	100
Modulus	N/mm^2		3.8
Limiting oxygen index (LOI)	%		62

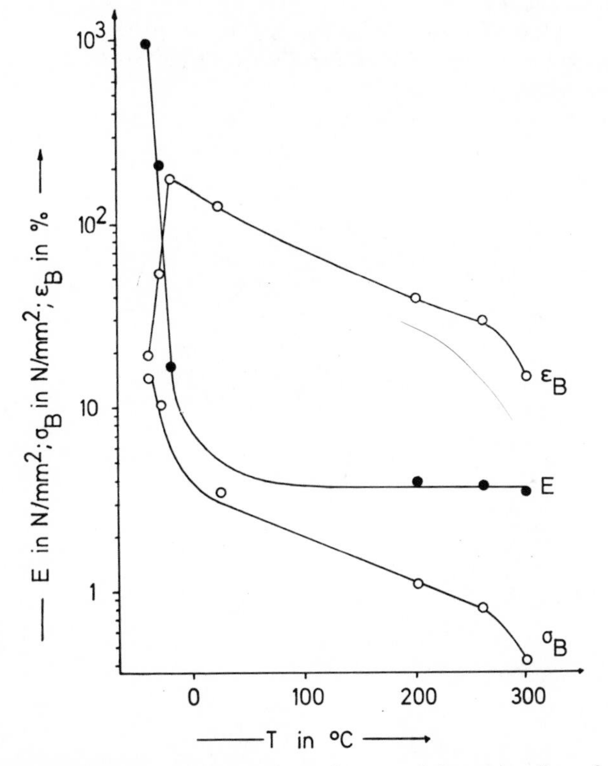

Figure 12-3 Tensile strength σ_B, ultimate elongation ε_B and elongation E as a function of temperature for the same polymer as in Figure 12-2 (according to data from Ref. 3).

tions of Dexsil 202 in xylene can be used for instance in the formation of coatings on metals, glass and fabrics, for example, for electrical parts and for autoclaves. A special type, Dexsil 300-GC is suggested as the liquid phase for the gas chromatographic separation of high boiling materials.

Poly(carborane siloxanes) are barely swollen by low molecular weight alcohols. But they do swell more than the silicones in hydrocarbons and ketones. On the other hand, they are significantly more stable to hydrolysis than the silicone rubbers.

References

1. H. A. Schroeder, O. G. Schaffling, T. B. Larchan, F. F. Trulla and H. L. Heying, *Rubber Chem. Tech.*, **39**, 1184 (1966).
2. Anonym., *Chem. Eng. News*, **49**(12), 46 (March 22, 1971).
3. E. N. Peters, E. Hedaya, J. H. Kawakami, G. T. Kwiatokowski, D. W. McNeill and R. W. Tulis, *Rubber Chem. Tech.*, **48**, 14 (1975).
4. Anonym., *Kunststoffe*, **63**, 616 (1973).
5. M. B. Roller and J. K. Gillham, *Polymer Eng. Sci.*, **14**, 567 (1974).
6. M. B. Roller and J. K. Gillham, *J. Appl. Polymer Sci.*, **17**, 2141 (1973).

13 Modified Organic Natural Materials

13.1 HYDROXYPROPYLCELLULOSE

13.1.1 Structure and Synthesis

Hydroxypropylcellulose with the idealized structure

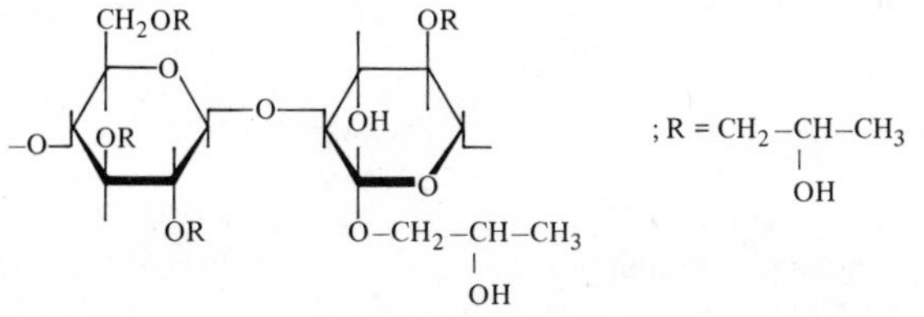

CH_2OR, OR, O, –O–, OR, OH, O, OR, $O{-}CH_2{-}CH(OH){-}CH_3$

; $R = CH_2{-}CH(OH){-}CH_3$

results from the reaction of alkali cellulose with propylene oxide at elevated temperatures and pressures.[1,2] Production was announced in 1968 and begun in 1969.[3] Hercules Inc. manufactures a standard type for industrial and cosmetic use under the tradename Klucel® and a specialty type for food and pharmaceutical use under the name Klucel F.

Practically all the primary hydroxyl groups undergo ether formation; the secondary cellulose hydroxyl groups do not. A second propylene oxide can react with the newly formed hydroxyl of the isopropyl ether; consequently, the degree of substitution on the hydroxyl groups must be differentiated from their degree of reaction. The degree of molecular substitution, MS, represents the mean number of reacted propylene oxide molecules *per* anhydro glucose residue. In the structure above it is three. The average degree of substitution, DS, which cannot be reliably determined experimentally, describes the number of substituted hydroxyl groups *per* anhydro glucose residue. In the case above, DS = 3 for the left anhydro glucose unit and DS = 2 for the right one.

13.1.2 Properties and Application

The hydroxypropylcelluloses HPC are soluble in water at temperatures under 38 °C but not above.[4-6] They are insoluble in 10 % aqueous sodium chloride, and they are the only commercial, water-soluble cellulose derivatives which are soluble in organic solvents such as dimethyl sulfoxide, dimethylformamide, acetic acid, pyridine, methanol and ethanol. They are swollen by benzene, carbon tetrachloride, ethylene glycol and glycerin. HPC's are insoluble in aliphatic hydrocarbons. Low molecular weight HPC's are soluble in acetone.

Films can be cast from solution. Table 13-1 lists the reported properties. HPC can also be processed by injection molding, extrusion and compression molding. Films and molded articles are stable to enzymatic degradation.

TABLE 13-1
Properties of hydroxypropylcellulose films.

Property	Physical unit	Values measured from Ref. 3	Ref. 7
Molecular weight (weight average)	g/mol		200,000
Degree of molecular substitution		4.0	
Density	g/cm^3	1.23	1.1
Tensile strength at fracture	N/mm^2	14–28	14
Modulus	N/mm^2		430
Ultimate elongation	%	56	50
Folding endurance flexibility		10,000	10,000
Gas permeability O_2	$cm^3\ s\ g^{-1}$		56×10^{-17}
N_2	$cm^3\ s\ g^{-1}$		7×10^{-17}
CO_2	$cm^3\ s\ g^{-1}$		383×10^{-17}
Water absorption (equilibrium, 23°C, 50 % rel. humidity)	%	2.5	4
Dielectric constant	1		9.07
Dissipation factor	1		0.0706
Volume resistivity	Ω cm		5×10^9

Typical applications are: aerosol stabilizers and film formers, thickeners for adhesive materials, binders, cosmetics, pharmaceuticals, printing inks and cleaners, oil-impermeable coatings in the paper and food industries and protective colloids for suspension polymerization.[7]

References

1. U.S. Patent 3278520 (February 8, 1963); Hercules Inc.; inv.: E. D. Klug; *C. A.*, **62**, 7993h (1965).
2. U.S. Patent 3278521 (February 8, 1963); Hercules Inc.; inv.: E. D. Klug; *C. A.*, **62**, 7993e (1965).
3. E. D. Klug, *Food Technol.*, **24**, 51 (1970).
4. J. H. Elliott, *J. Appl. Polymer Sci.*, **13**, 755 (1969).
5. R. J. Samuels, *J. Polymer Sci.* [*A-2*], **7**, 1197 (1969).
6. M. G. Wirich and M. H. Waldman, *J. Appl. Polymer Sci.*; **14**, 579 (1970).
7. Product information literature of the firm Hercules Inc.

13.2 HYDROXYPROPYLMETHYLCELLULOSE

The Methocel®-series of Dow Chemical Co.[1] are cellulose ethers of varying composition and viscosities. Methocel J, an hydroxypropylmethylcellulose of undisclosed DS and MS values (see Section 13.1.1), is new on the market. Its density is 1.39 g/cm^3. Methocel J is water-soluble and enzyme-resistant. It is offered in three different viscosity grades. Typical applications are as thickeners, emulsifiers and stabilizers for emulsion dyes and coatings.

Reference

1. Product information literature of the Dow Chemical Co.

13.3 HYDROXYETHYLMETHYLCELLULOSE

Dow Chemical[1] produces in research quantities an hydroxyethylmethylcellulose under the designation "Experimental Cellulosic Thickener XD-7630" which was developed as a thickener for paints. The content and distribution of substituents were not disclosed. The newly developed product is stable to enzymatic hydrolysis and does not gel thermally. Aqueous solutions have a higher surface tension and, therefore, foam less.

Reference

1. Product information literature of the Dow Chemical Co.

13.4 HYDROXYALKYLGUARAN

The guar plant is grown in India, Pakistan and in the southwestern United States. Its seeds yield the polysaccharide guaran (called guar gum, or simply guar, in the U.S.). Guaran has a molecular weight of about 200,000 g/mol. Every second sugar residue of the poly(β-(1→4)-mannopyranosyl) main chain has a single D-galactose unit bound through α-(1→6) as a side group.

Stein-Hall (a subsidiary company of Celanese) introduced in 1972 a hydroxyethylated guar under the tradename Jaguar HE and a hydroxypropylated guar under the name Jaguar HP. According to company literature,[1] the positions marked with * in the structural formula below are substituted with $(CH_2CH_2O)_nH$ (in Jaguar HE) and with $(CH_2CH(CH_3)O)_nH$ (in Jaguar HP). The DS- or MS-values (see Section 13.1.1) have not been revealed.

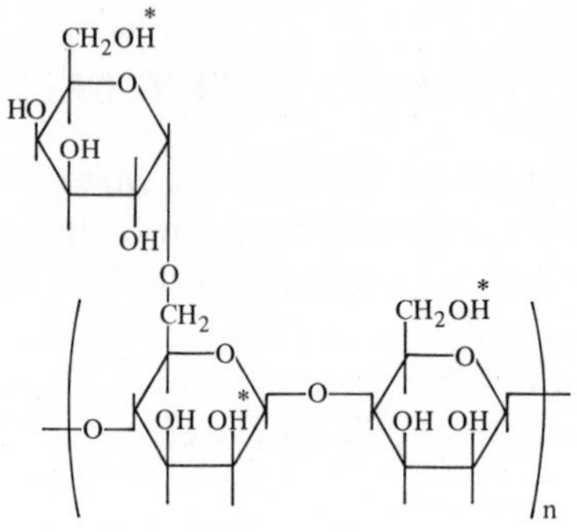

Flocculating tendency is reduced or even eliminated by the hydroxyalkylation. During solution of the guar, the viscosity both increases and reaches its final value faster for the hydroxyalkylated- than for the non-hydroxyalkylated-guar. Commercial literature calls this behavior "faster hydration". The viscosity of aqueous solutions is constant for over 24 h at 60°C, and almost constant for over 2 h at 80°C. It is dependent on the shear rate, as are all non-Newtonian liquids, and is independent of pH in the range pH = 2–11.5. Addition of multivalent ions causes gelling. Glossy, clear films can be cast from solution. The polymers are biologically degraded more easily than the cellulose ethers.

Typical applications of the hydroxyalkylated guars include use as thickening agents, strength improvers for paper, for the mining industry and sizing in the textile industry.

Reference

1. Product information literature of the firm Stein-Hall.

13.5 CHITOSAN

Chitin is poly(β-(1→4)-*N*-acetyl-2-amino-2-desoxyglucopyranose):

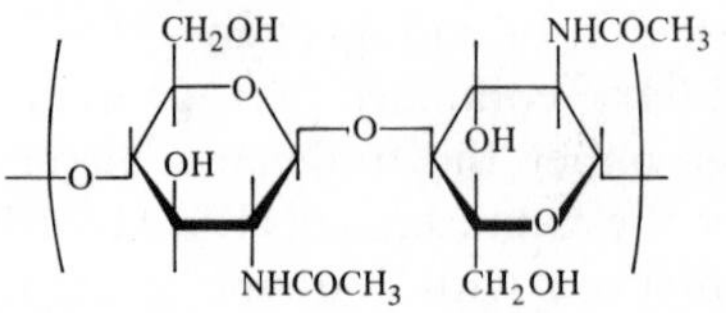

It is the structural substance of crustacea in which it is associated with calcium carbonate (about 70%). Treatment of the shells with 5% hydrochloric acid releases the chitin. Saponification (deacetylation) with sodium hydroxide gives chitosan.

Of the various suggested uses for chitosan, the following appear to be especially promising: biologically degradable films for food packaging, additives to improve the wet strength of paper, ion-exchange resins for water purification and as an aid for more rapid healing of wounds.[1] Patents covering chitosan-production have recently been acquired by Marine Commodities International.[2]

References

1. Anonym., *Chemical Week*, **116**(17), 33 (April 23, 1975).
2. Private communication, Marine Commodities International.

13.6 GRAFTED VISCOSE FIBERS

13.6.1 Structure and Synthesis

Although grafting on cellulose fibers was thoroughly investigated a long time ago (see Ref. 1), it never succeeded commercially for a number of reasons. The graft reactions were inefficient and they lacked specificity; the actual improvements were trivial; harmful side products appeared, and the costs were high. Still, a procedure[2] developed by Scott Paper Co. appears promising. It depends on a thiocarbonate/peroxide-redox process,[3–5] and it is suitable for the manufacture of flame-retardant cellulose fibers[6] and for dispersable viscose fibers for making nonwoven fabrics.[7] Both products are made in experimental quantities by the firm Chemiefaser Lenzing.[8]

In the procedure, the cellulose to be grafted is first partially converted to the corresponding xanthate.

$$\text{cell–CH}_2\text{OH} + \text{CS}_2 + \text{NaOH} \longrightarrow \text{cell-CH}_2\text{O–}\underset{\underset{\text{S}}{\|}}{\text{C}}\text{–SNa} + \text{H}_2\text{O} \qquad (13\text{–}1)$$

and then with iron salts to its corresponding iron compound, which in turn reacts with hydroxy radicals (from hydrogen peroxide decomposition) to form radical sites on the cellulose:

$$\text{cell–CH}_2\text{–O–}\underset{\underset{\text{S}}{\|}}{\text{C}}\text{–S}\cdots\text{Fe} + \text{HO}^\bullet \longrightarrow \text{cell–}\overset{\bullet}{\text{C}}\text{H–O–}\underset{\underset{\text{S}}{\|}}{\text{C}}\text{–S}\cdots\text{Fe} + \text{H}_2\text{O} \qquad (13\text{–}2)$$

$$\text{cell–CH}_2\text{–O–}\underset{\underset{\text{S}}{\|}}{\text{C}}\text{–SH} + \text{HO}^\bullet \longrightarrow \text{cell–CH}_2\text{–O–}\underset{\underset{\text{S}}{\|}}{\text{C}}\text{–S}^\bullet + \text{H}_2\text{O} \qquad (13\text{–}3)$$

The cellulose radicals in principle can also arise from transfer from the hydroxyl radicals and/or growing polymer radicals. All these radicals can initiate vinyl- and acrylic monomer polymerization. for instance

$$\text{cell–CH}_2\text{–O–}\underset{\underset{\text{S}}{\|}}{\text{C}}\text{–S}^\bullet + \text{CH}_2\text{=CHX} \longrightarrow \text{cell–CH}_2\text{–O–}\underset{\underset{\text{S}}{\|}}{\text{C}}\text{–S–CH}_2\text{–}\overset{\bullet}{\text{C}}\text{HX} \qquad (13\text{–}4)$$

The exact nature of the grafted, phosphorus-containing vinyl monomers has not been published, although the following types of compounds are discussed:[6]

$$\text{CH}_2\text{=CH—O—}\underset{\underset{\text{O}}{\|}}{\text{P}}\text{(OR)}_2 \qquad \text{CH}_2\text{=CH—CO—NH—R—O—}\underset{\underset{\text{O}}{\|}}{\text{P}}\text{(OR')}_2$$

No data are available concerning the grafted monomer for the production of dispersible viscose fibers for nonwoven fabrics.

13.6.2 Properties and Application

Practically no homopolymers form during the grafting. Grafted side-chain length is relatively short: according to hydrolytic degradation of the cellulose chain, the side branches' molecular weights are about 30,000 to 120,000 g/mol.[4] According to measurements on the titer distribution before and after grafting, grafting takes place uniformly and without noticeable change in fiber length.[3] However, the fiber cross-section does increase.

Because of the increase in cross-section, tensile strength with respect to titer falls off almost directly with the extent of grafting (Figure 13-1), even though absolute tensile strength after grafting is slightly increased. Therefore, in order to produce graft-modified cellulose fibers with sufficient strength, one must start out with high-strength starting fibers. The difference in tenacity between conditioned and wet fibers shown in Figure 13-1 is immaterial for high-modulus fibers. Loop tenacity and knot strength are also reduced by the grafting, although to a lesser extent than the tensile strength. Both appear to be dependent on the chemical nature of the grafted polymer.

The viscose fiber becomes flame retardant at a phosphorus content of 3%.[5] With a combination of phosphorus- and nitrogen-containing compounds 2% is sufficient. The flame retardant finish remains intact even after fifty washings.[5,6] Dying properties are unaffected. Somewhat problematical, however, is the fabrication of the grafted staple fiber in yarn spinning, since the grafted products tend to stick together.

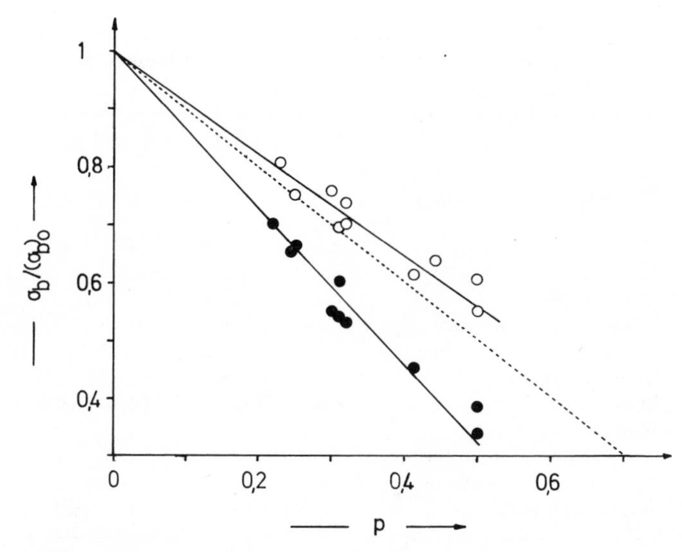

Figure 13-1 Decrease in the tenacity with respect to titer, $\sigma_b/(\sigma_b)_o$, as a function of the proportion p of grafted monomer for normal, conditioned viscose staple fibers (○) and normal, wet viscose staple fibers (●) (according to Ref. 3).

References

1. H. Krässig and V. Stannett, *Adv. Polymer Sci.*, **4**, 111 (1965).
2. U.S. Patent 3359224 (December 19, 1967); Scott Paper Co.; inv.: R. W. Faessinger and J. S. Conte; *C. A.*, **68**, 41239 (1968).
3. H. Krässig, *Das Papier*, **24**, 926 (1970).
4. H. Krässig, *Svensk Papperstidning*, **74**, 417 (1971).
5. H. E. Teichmann, W. J. Brickman, R. W. Faessinger, C. G. Mayer and H. A. Krässig, *Tappi*, **57**(7), 61 (1974).
6. J. Harms and H. Krässig, *Lenzinger Berichte*, **36**, 214 (1974).
7. F. Gotschy and H. Krässig, *Das Papier*, **26**, 813 (1972).
8. H. Krässig, private communication.

13.7 GRAFTED CASEIN FIBERS

A silk-like fiber is produced by the firm Toyobo from the radical grafting of 70% acrylonitrile on 30% casein followed by wet spinning. It was marketed first under the tradename K-6-fiber, then as "Chinon".[1] The silk-like properties are evident on comparison with the properties of natural silk (Table 13-2). Elastic recovery and stability to sunlight are far better than natural silk's, while dyeability and tendency to electrostatic charge build-up are about the same (and, therefore, far better than in synthetic fibers).

TABLE 13-2
Comparison of the properties of Chinon-fiber with those of natural silk (according to data in Ref. 1).

Property	Physical unit	Values measured for	
		Chinon	Natural silk
Density	g/cm^3	1.20	1.33–1.45
Tenacity, dry	km	36–47	27–36
wet	km	32–43	19–25
Ultimate elongation, dry	%	15–25	15–25
wet	%	16–27	27–33
Knot strength	km	9–18	26
Modulus	N/mm^2	3900–9800	6400–11,800
Water absorption (20°C, 60% relative humidity)	%	5–6	9
Shrinkage in boiling water	%	2.5–4.5	0–1
Surface resistivity		8×10^{10}	1.5×10^{10}

Reference

1. S. Morimoto, *Ind. Eng. Chem.*, **62**(3), 23 (1970).

14 Outlook

The foregoing chapters describe over fifty polymers with new chemical structures, which were either ready for the market or on the market between 1969 and 1975. The extent of their novelty ranges from one extreme to the other. At the one end are the new derivatives of time-honored cellulose, at the other end are polymers with completely new structural elements, such as poly(terephthaloyloxamidrazone) or carboxynitroso-rubber. In between are the new polymers with familiar functional groups but with new base units such as the polyamides and the polyether esters and the many new copolymers. The possibilities for synthetic macromolecular chemistry are far from exhausted at this point. In all likelihood the next five years will witness the appearance of additional polymeric materials with new chemical structures, even if not in such abundance.

One can clearly distinguish several classifications among the polymers which have been described. There is considerable activity in the area of flame-retardant fibers and plastics. Certain new plastics are gaining ever increasing importance as engineering materials. For the elastomers, research efforts appear to be directed principally towards two aims: new all-purpose rubbers and rubbers with improved methods of fabrication (powder- or liquid-form rubbers). A trend towards fast curing, solvent-free systems is unmistakable in films and coatings.

The new developments in the polyimide area merit attention as far as chemical structure is concerned. The activity in the area of sulfur-containing polymers and fluoropolymers is striking. Undoubtedly, increased numbers of polymers with heterocyclic groups will come on the market in the next few years.

The overwhelming majority of polymers introduced onto the market are organic; most of these based on petroleum products. Consequently it is often asked whether long-term supply of petrochemical raw materials to the polymer industry is assured.

Apparently the answer to this question is "yes" (see, for instance, Ref. 1). At the moment only 5–6% of petroleum is converted into petrochemical products, the majority of the rest is used as energy carriers (gasoline, heating oil). The opening-up of economically feasible energy sources would free considerable amounts of petroleum for the petrochemical industry, providing the supply of petroleum remains constant. In the United States approximately

40% of the petroleum is converted into gasoline. Reduction in gasoline consumption by 15% (that is from 40 to 34%) would practically cover the whole petrochemical requirement of the United States. Because petrochemical products have a greater value than does gasoline, a shift in this direction for purely economic considerations will undoubtedly take place. It is already evident in the United States, when one notes the increase in the number of refineries working with naphtha.[2]

Oil from oil shale and oil sand will not yet reach commercial importance in the next decade for economic reasons. At the moment revival of coal technology to produce basic materials for the polymer industry appears pretty unlikely. Actually it seems more sensible to use coal as an energy source and oil as a chemical-product source, and not the reverse, since the complicated chemical synthesis from coal requires extensive separation and purification operations. Especially in the United States, the demand is being made for renewable raw materials. The only possible truly renewable raw material for organic polymers is, of course, carbon dioxide. Carbon dioxide appears in dilute form in the atmosphere and in concentrated form in calcium carbonate. But obtaining carbon dioxide from these two sources requires so much energy that such processes are economically unfeasible for the time being.

A return to the naturally occurring conversion products of carbon dioxide, that is plant and animal raw materials, is technically and economically possible. The manufacture of polymers and polymer products out of animal raw materials has stagnated for quite some time (see Figure 1-2); except for chitosan (Chapter 13.5) no new polymeric materials derived from animals have come onto the market (see Table 14-1). But animal substances have to be discounted as a high volume source of raw materials because the animals themselves live off plant-life (and other animal-life). Many raw materials can, therefore, be more easily and directly obtained from plant products.

TABLE 14-1
Polymers derived from animal raw materials.

Source of raw material	Polymer or polymeric product
Sheep, llama etc.	wool
Silk worms	natural silk
Animal skins	leather, genuine parchment, collagen, gelatins
Milk	casein
Crustacae	chitosan

A large number of polymers and polymeric products are manufactured out of plant materials (Table 14-2). The polyamides and diallyl monomers based on crambe stand out among the new, suggested but not yet commercialized products. Although production of several of these polyamides appears economically feasible, its realization does suffer from problems common to all raw materials from plant sources. These are that raw material supply is not assured (political influences, weather), occurrence of variable quality from year to year, or that limited farm land must be sacrificed. Therefore, the use of plant waste material would be desirable. The use of waste cellulose appears particularly promising (see Ref. 1), and in particular production of glucose as a possible raw material for the polymer industry. Fermentation of glucose to α-amino acids is already being carried out on a commercial scale in several countries. Glutamic acid costs about 1 $/kg at the present time. Therefore, poly(α-amino acid) polymers (Chapter 8.1) could easily be important as fibers, films and synthetic leather in the future.

Apart from a few exceptions, the future of plant materials appears to be less in the direct utilization of naturally occurring polymers than as a source for intermediate products. Natural polymers ultimately are not synthesized from plants merely to be used as truck tires (natural rubber) or men's underwear (cotton). For the most part they are not optimal for the intended applications.

TABLE 14-2
Polymers derived from plant raw materials.

Raw material source	Intermediate or monomer	Polymer	Ref.
Commercialized:			
Natural rubber		natural rubber	3
Balata, gutta-percha		balata, gutta-percha	3
Chicle		chicle	3
Cotton, hemp, flax, etc.		cotton, hemp, linen, etc.	3
Algae, seeds and other plant materials		vegetable gums	4
Wood		cellulose fiber and other cellulose derivatives	3
Cellulose	glucose,	dextran, xanthan	5
	α-amino acids	poly(glutamate)	1
Starch		amylose, amylopectin and their derivatives	4
Castor oil	ricinoleic acid	polyamide 11 and 6, 10	3
Corn cobs	tetrahydrofuran	poly(tetrahydrofuran)	3
	hexamethylenediamine	polyamide 6,6	3
Proposed, but not commercialized:			
Soybean	azelaic aldehyde soyanitrile	polyamide 9	6, 7
	ditto	polyamide 6,9; 9,9 and 9.12	7
Crambe	erucic acid or brassylic acid	polyamide 13 and 6,13	8
	ditto	polyamide 13,13	9–11
	ditto	poly(diallyl brassylate)	12
Starch		pullulan	13

Natural polymers do have one property which most synthetic polymers lack: they are part of an ecosystem which has evolved over thousands of years and, therefore, are biodegradable. The biological degradability requirement was aimed principally at packaging materials made from synthetic polymers, while materials such as glass and metals did not come under nearly as much fire. This, despite the fact that they do not decompose rapidly either, which by itself shows that the demand for biologically degradable polymers is really irrational. Fundamentally it is not even a problem of biological contamination of the environment, because, as a rule, polymers can neither be directly metabolized nor reabsorbed. Really it is an esthetic problem (refuse does not look good), which should be solved by means of education. In fact, it it is to be feared that some of the newly developed biologically degradable polymers are going to do just what they should not—pollute the environment. Biological degradation of these newly developed polymers[14-16] produces oligomers,[16] which, in contrast to the polymers themselves, can be easily metabolized by microorganisms. There are no published systematic investigations on environmental pollution by oligomers (see Refs. 17 and 18). Apparently the danger of such pollution by biologically degradable poly(alkanes) is not too great because certain known microorganisms can transform hydrocarbons into proteins. Doubts persist, on the other hand, concerning halogenated and aromatic compounds. However, political pressure will probably give certain priority to the search for biologically degradable polymers.

References

1. H.-G. Elias, *Chem. Tech.*, **5**, 748 (1975); **6**, 244 (1976).
2. Anonym., *Modern Plastics*, **51**(11), 21 (1974).
3. H.-G. Elias, *Makromolekule*, Hüthig und Wepf, Basel, 3rd German ed., 1975; *Macromolecules*, Plenum Press, New York, 1977.
4. R. L. Whistler, *Industrial Gums*, Academic Press, New York, 1973.
5. A. Jeanes, *J. Polymer Sci.* [*Symp.*], **C45**, 209 (1974).
6. W. R. Miller, E. H. Pryde, R. A. Awl, W. L. Kohlhase and D. J. Moore, *Ind. Eng. Chem. Prod. Res. Dev.*, **10**, 442 (1971).
7. R. B. Perkins, J. J. Roden, III and E. H. Pryde, *J. Amer. Oil Chem. Soc.*, **52**, 473, (1975).
8. J. L. Greene, Jr., R. E. Burke, Jr. and I. A. Wolff, *Ind. Eng. Chem. Prod. Res. Dev.*, **8**, 171 (1969).
9. J. L. Greene, Jr., E. L. Huffman, R. E. Burke, Jr. and W. C. Sheehan, *J. Polymer Sci.* [*A-1*], **5**, 391 (1967).
10. R. B. Perkins, J. J. Roden, III, A. C. Tanquary and J. A. Wolff, *Modern Plastics*, **46**(5), 136 (1969).
11. Anonym., *Chem. Tech.*, **2**, 515 (1972).
12. S.-P. Chang, T. K. Miwa and W. H. Tallent, *J. Appl. Polymer Sci.*, **18**, 319 (1972).
13. Anonym., *Chem. Eng. News*, **51**, 40 (Dec. 24, 1973).
14. F. Rodriguez, *Chem. Tech.*, **1**, 409 (1971).

15. G. Scott, *Europ. Polymer J.* [*Suppl.*], **1969**, 189; *Kunststoff-Rdsch.*, **17**, 548 (1969).
16. E. Dan and J. E. Guillet, *Macromolecules*, **6**, 230 (1973).
17. J. A. Zapp, *Arch. Environm. Health*, **4**(3), 335 (1962).
18. F. Bischoff, *Clin. Chem.*, **18**, 869 (1972).

Appendix

TABLE A-1

Internationally used abbreviations for thermoplastics, thermosets, fibers, elastomers, and additives.

(According to DIN 7723 and 7728; ASTM D 1600-64 T and 1418-67; BS 3502-1962; ISO/DR 1252; IUPAC; EEC.)

ABR	Poly(acrylic ester-co-butadiene) [elastomer; ASTM], see also AR
ABS	Poly(acrylonitrile-co-butadiene-co-styrene) [ASTM; DIN; ISO]
ACM	Poly(acrylic ester-co-2-chlorovinyl ether) [elastomer; ASTM]
ACS	Blend of poly(acrylonitrile-co-styrene) with chlorinated poly(ethylene)
AFMU	Poly(tetrafluoroethylene-co-trifluoronitrosomethane-co-nitrosoperfluorobutyric acid) = Nitroso rubber [ASTM]
AMMA	Poly(acrylonitrile-co-methyl methacrylate) [DIN; ISO]
ANM	Poly(acrylonitrile-co-acrylic ester) [elastomer; ASTM]
AP	Poly(ethylene-co-propylene) [elastomer], see also APK, EPM, and EPR
APK	Poly(ethylene-co-propylene) [elastomer], see also AP, APT, EPM, and EPR
APT	Poly(ethylene-co-propylene-co-diene) [elastomer], so-called ethylene-propylene terpolymer, see also EPDM, EPT, and EPTR
AR	Acrylic ester-elastomer, see also ABR, ACM, ANM
ASA	Poly(acrylonitrile-co-styrene-co-acrylic ester) [DIN]
ASE	Alkyl sulfonic acid ester [ISO]
AU	Polyurethane elastomer with polyester segments [ASTM]
BBP	Benzyl butyl phthalate [DIN; ISO]
BOA	Benzyl octyl adipate [ISO]
BR	Poly(butadiene) [elastomer; ASTM]
BT	Poly(butene-1)
Butyl	Poly(isobutylene-co-isoprene) [so-called butyl rubber; BS]
CA	Cellulose acetate [ASTM; DIN; ISO]
CAB	Cellulose acetobutyrate [ASTM; DIN; ISO]
CAP	Cellulose acetopropionate [ASTM; DIN]
CAR	Carbon fiber
CF	Cresol-formaldehyde resin [DIN]
CFK	Man-made fiber reinforced plastics
CFM	Poly(trifluorochloroethylene) [ASTM], see also PCTFE
CHC	Poly(epichlorohydrin-co-ethylene oxide) [elastomer], see also CHR, CO and ECO
CHR	Poly(epichlorohydrin) [elastomer], see also CHC, CO, and ECO
CL	Poly(vinyl chloride) fiber [EEC]
CM	Chlorinated poly(ethylene) [ASTM], see also CPE

(Table A-1, *cont.*)

CMC	Carboxymethyl cellulose [ASTM; DIN]
CN	Cellulose nitrate [ASTM; DIN], see NC
CNR	Carboxy nitroso rubber, see also AFMU
CO	Poly(epichlorohydrin) = "polychloromethyloxirane" [elastomer; ASTM], see also CHC, CHR, and ECO
CP	Cellulose propionate [DIN; ISO]
CPE	Chlorinated poly(ethylene), see also CM
CPVC	Chlorinated poly(vinyl chloride), see PC, PeCe, and PVCC
CR	Poly(chloroprene) [elastomer; ASTM, BS]
CS	Casein–formaldehyde resin
CSM	Chlorosulfonated poly(ethylene) [ASTM], see CSPR, CSR
CSPR	Chlorosulfonated poly(ethylene) [BS], see CSM, CSR
CSR	Chlorosulfonated poly(ethylene)
CTA	Cellulose triacetate
DABCO	Triethylenediamine
DAP	Diallyl phthalate (resins) [ASTM; DIN], see FDAP
DBP	Dibutyl phthalate [DIN, ISO, IUPAC]
DCP	Dicapryl phthalate [DIN, ISO, IUPAC]
DDP	Didecyl phthalate
DEP	Diethyl phthalate [ISO]
DHP	Diheptyl phthalate [ISO]
DHXP	Dihexyl phthalate [ISO]
DIBP	Diisobutyl phthalate [DIN, ISO]
DIDA	Diisodecyl adipate [DIN, ISO, IUPAC]
DIDP	Diisodecyl phthalate [DIN, ISO, IUPAC]
DINA	Diisononyl adipate [ISO]
DINP	Diisononyl phthalate [DIN, ISO]
DIOA	Diisooctyl adipate [DIN, ISO, IUPAC]
DIOP	Diisooctyl phthalate [DIN, ISO, IUPAC]
DIPP	Diisopentyl phthalate
DITDP	Diisotridecyl phthalate [DIN, ISO], see DITP
DITP	Diisotridecyl phthalate [DIN], see DITDP
DMF	Dimethylformamide
DMP	Dimethyl phthalate [ISO]
DMT	Dimethyl terephthalate
DNP	Dinonyl phthalate [ISO, IUPAC]
DOA	Dioctyl adipate, di-2-ethylhexyl adipate [DIN, ISO, IUPAC]
DODP	Dioctyldecyl phthalate [ISO], see ODP
DOIP	Dioctyl isophthalate, di-2-ethylhexyl isophthalate [DIN, ISO]
DOP	Dioctyl phthalate, di-2-ethylhexyl phthalate [DIN, ISO, IUPAC]
DOS	Dioctyl sebacate, di-2-ethylhexyl sebacate [DIN, ISO, IUPAC]
DOTP	Dioctyl terephthalate, di-2-ethylhexyl terephthalate [DIN, ISO]
DOZ	Dioctyl azelate, di-2-ethylhexyl azelate [DIN, ISO, IUPAC]
DPCF	Diphenyl cresyl phosphate [ISO]
DPOF	Diphenyl octyl phosphate [ISO]
DUP	Diundecyl phthalate

(Table A-1, *cont.*)

EA	Segmented polyurethane fibers
EC	Ethyl cellulose [DIN]
ECB	Blends of ethylene copolymers with bitumen
ECO	Poly(epichlorohydrin) [elastomer; ASTM], see CHC, CHR, and CO
EEA	Poly(ethylene-co-ethyl acrylate) [ISO]
ELO	Epoxidized linseed oil
EP	Epoxide resin
EPDM	Poly(ethylene-co-propylene-co-diene) [elastomer], see also APT, EPT, and EPTR
EP-G-G	Epoxide resin-textile glass fabric prepreg
EP-K-L	Epoxide resin carbon fiber fabric prepreg
EPM	Poly(ethylene-co-propylene) [elastomer; ASTM, ISO], see AP, APK, EPR
EPR	Poly(ethylene-co-propylene) [elastomer; BS], see AP, APK, EPM
EPS	Poly(styrene) foam
EPT	Poly(ethylene-co-propylene-co-diene) [elastomer], see APT, EPDM, EPTR
EPTR	Poly(ethylene-co-propylene-co-diene) [elastomer; BS], see APT, EPDM, and EPT
E-PVC	Emulsion PVC
E-SBR	Emulsion SBR
ESO	Epoxidized soybean oil [DIN, ISO]
ETFE	Poly(ethylene-co-tetrafluoroethylene)
EU	Polyurethane elastomers with polyether segments [ASTM]
EVA	Poly(ethylene-co-vinyl acetate) [DIN, ISO]
EVAC	Poly(ethylene-co-vinyl acetate) [elastomer]
FDAP	Diallyl phthalate(resins), see also DAP
FE	Fluoro-containing elastomers
FEP	Poly(tetrafluoroethylene-co-hexafluoropropylene) [DIN, ISO], see PFEP
FPM	Poly(vinylidene fluoride-co-hexafluoropropylene) [ASTM]
FSI	Fluorosilicones [ASTM]
GEP	Glass fiber reinforced epoxide resin
GF	Glass fiber reinforced plastics, see also GFK, RP
GF-EP	Glass fiber reinforced epoxide resins
GFK	Glass fiber reinforced plastics, see also GF, RP
GF-PF	Glass fiber reinforced phenolic resins
GF-UP	Glass fiber reinforced unsaturated polyester resins
GR-I	Old U.S. name for butyl rubber
GR-N	Old U.S. name for nitrile rubber
GR-S	Old U.S. name for styrene/butadiene rubber
GUP	Glass fiber reinforced unsaturated polyester resin
GV	Glass fiber reinforced thermoplastics
HDPE	High-density poly(ethylene); in German literature sometimes also for high pressure PE (i.e. with low density)
HMWPE	Unbranched poly(ethylene) with very high molecular weight
HPC	Hydroxypropyl cellulose

(Table A-1, *cont.*)

IIR	Poly(isobutylene-co-isoprene) [elastomer; ASTM], see butyl, PIB, GR-I
IR	*cis*-1,4-poly(isoprene), synthetic [ASTM, BS]
KFK	Carbon fiber reinforced plastics [DIN]
LDPE	Low-density poly(ethylene)
L-SBR	Solution-polymerized SBR
MA	Modacrylic fiber
MBS	Poly(methyl methacrylate-co-butadiene-co-styrene)
MC	Methyl cellulose
MDI	4,4-diphenylmethanediisocyanate
MDPE	Medium-density poly(ethylene) (*ca.* 0.93–0.94 g/cm^3)
MF	Melamine-formaldehyde resin [ASTM, DIN, ISO]
MFK	Metal-fiber reinforced plastics
MOD	Modacrylic fiber [EEC]
MP	Melamine–phenol–formaldehyde resins
M-PVC	Bulk polymerized PVC
NBR	Poly(butadiene-co-acrylonitrile), nitrile rubber [ASTM, BS], see PBAN
NC	Cellulose nitrate, see CN
NCR	Poly(acrylonitrile-co-chloroprene) [ASTM]
NDPE	Low density poly(ethylene), see LDPE
NK	Natural rubber, see NR
NR	Natural rubber [ASTM], see NK
ODP	Octyldecyl phthalate [ISO], see DODP
OER	Oil-extended rubber
PA	Polyamide [ASTM, DIN, ISO]: the first number refers to the number of methylene groups in aliphatic diamines; the second number gives the number of carbon atoms in the aliphatic dicarboxylic acid; I stands for isophthalic acid, T for terephthalic acid. A single number refers to the polyamide of an α,ω-amino acid (or its lactam).
PAA	Poly(acrylic acid)
PAC	Poly(acrylonitrile) [IUPAC], see PAN, PC
PAN	Poly(acrylonitrile), see PAC, PC (also as trademark)
PB	Poly(butene-1) [DIN], see PBT
PBAN	Poly(butadiene-co-acrylonitrile) [elastomer]
PBR	Poly(butadiene-co-pyridine) [ASTM]
PBS	Poly(butadiene-co-styrene), see SBR
PBT	Poly(butene-1), see PBT
PBTP	Poly(butylene terephthalate) [DIN], see PTMT
PC	1) Polycarbonate[ASTM, DIN, ISO] 2) Poly(acrylonitrile) [EEC], see PAC, PAN 3) Previously: postchlorinated PVC
PCF	Poly(trifluorochloroethylene) fiber
PCTFE	Poly(trifluorochloroethylene) [DIN], see CFM

(Table A-1, *cont.*)

PCU	Poly(vinyl chloride)
PDAP	Poly(diallyl phthalate) [DIN], see DAP, FDAP
PE	1) Poly(ethylene)[ASTM, DIN, ISO] 2) Polyester fiber [EEC]
PEC	Chlorinated poly(ethylene)[DIN], see CPE
PeCe	Chlorinated PVC, see CPVC, PC, PVCC
PEO	Poly(ethylene oxide), see PEOX
PEOX	Poly(ethylene oxide), see PEO
PES	1) Polyester fiber 2) Polyether sulfone
PET	Poly(ethylene terephthalate), see PETP
PETP	Poly(ethylene terephthalate)[ASTM, DIN, ISO], see PET
PF	Phenol–formaldehyde resin [ASTM, DIN, ISO]
PFEP	Poly(tetrafluoroethylene-co-hexafluoropropylene)
PI	*trans*-1,4-poly(isoprene), Guttapercha [BS]
PIB	Poly(isobutylene)[BS, DIN]
PIBI	Poly(isobutylene-co-isoprene), butyl rubber, see butyl, IIR
PIP	*cis*-1,4-poly(isoprene), synthetic
PL	Poly(ethylene)[EEC]
PMCA	Poly(methyl α-chloromethacrylate)
PMI	Poly(methacrylimide)
PMMA	Poly(methyl methacrylate)[ASTM, DIN, ISO]
PMP	Poly(4-methylpentene-1)[DIN]
PO	1) Poly(propylene oxide)[elastomer, ASTM] 2) Poly(olefins) 3) Phenoxy resins
POM	Poly(oxymethylene resins)[DIN, ISO]
POR	Poly(propylene oxide-co-allyl glycidyl ether)[elastomer]
PP	Poly(propylene)[ASTM, DIN, ISO]
PPO	Poly(phenylene oxide); also a registered trademark
PPSU	Poly(phenylene sulfone)[ISO], see PSU
PS	Poly(styrene)
PSAN	Poly(styrene-co-acrylonitrile)[DIN], see SAN
PSAB	Poly(styrene-co-butadiene)[DIN], see SB
PSI	Poly(methylphenylsiloxane)[ASTM]
PST	Poly(styrene) fiber
PS-TSG	Poly(styrene) injection molded foam
PSU	Poly(phenylene sulfone), see PPSU
PTF	Poly(tetrafluoroethylene) fiber
PTFE	Poly(tetrafluoroethylene)[ASTM, DIN, ISO]
PTMT	Poly(tetramethylene terephthalate) = poly(butylene terephthalate), see P-BMT
PU	Polyurethane elastomers [BS]
PUA	Polyurea fiber
PUE	Segmented polyurethane fibers
PUR	Polyurethane [DIN, ISO]
PVA	1) Poly(vinyl acetate), see PVAC 2) Poly(vinyl alcohol), see PVAL 3) Poly(vinyl ether)

(Table A-1, *cont.*)

PVAC	Poly(vinyl acetate)[ASTM, DIN, ISO]
PVAL	Poly(vinyl alcohol)[ASTM, DIN, ISO]
PVB	Poly(vinyl butyral)[ASTM, DIN]
PVC	Poly(vinyl chloride)[ASTM, DIN, ISO]
PVCA	Poly(vinyl chloride-co-vinyl acetate)[DIN], see PVCAC
PVCC	Chlorinated PVC [DIN], see CPVC, PC, PeCe
PVDC	Poly(vinylidene chloride)[DIN, ISO]
PVDF	Poly(vinylidene fluoride)[DIN, ISO], see PVF_2
PVF	Poly(vinyl fluoride)
PVF_2	Poly(vinylidene fluoride), see PVDF
PVFM	Poly(vinyl formal)[DIN, ISO], see PVFO
PVFO	Poly(vinyl formal)[DIN], see PVFM
PVID	Poly(vinylidene cyanide)
PVK	Poly(vinylcarbazole)[DIN, ISO]
PVM	Poly(vinyl chloride-co-vinyl methyl ether)
PVP	Poly(vinylpyrrolidone)
PVSI	Poly(dimethylsiloxane) with phenyl- and vinyl-groups [ASTM]
PY	Unsaturated polyester resins [BS]
RF	Resorcinol–formaldehyde resins
RP	Reinforced plastics
SAN	Poly(styrene-co-acrylonitrile)[DIN, ISO], see PSAN
SB	High-impact poly(styrene)[DIN, ISO]
SBR	Poly(styrene-co-butadiene)[ASTM, BS]
SCR	Poly(styrene-co-chloroprene)[ASTM]
SI	1) Silicones [DIN]
	2) Poly(dimethylsiloxane)[ASTM]
SIR	1) Silicone rubber
	2) Poly(styrene-co-isoprene)[ASTM]
SMR	Standardized Malaysian rubber
SMS	Poly(styrene-co-α-methylstyrene) [DIN, ISO]
S-PVC	Suspension polymerized PVC
TC	Technical grade natural rubber
TCEF	Trichloroethyl phosphate [ISO]
TCF	Tricresyl phosphate [DIN, ISO], see TCP, TKP, TTP
TCP	Tricresyl phosphate [IUPAC], see TCF, TKP, TTP
TDI	Toluenediisocyanate
TIOTM	Triisooctyl trimellitate [DIN, ISO]
TKP	Tricresyl phosphate, see TCF, TCP, TTP
TOF	Trioctyl phosphate, tri(2-ethylhexyl)phosphate [DIN, ISO], see TOP
TOP	Trioctyl phosphate, tri(2-ethylhexyl)phosphate [IUPAC], see TOF
TOPM	Tetraoctyl pyromellitate [DIN, ISO]
TOTM	Trioctyl mellitate [DIN, ISO]
TPA	1,5-*trans*-poly(pentenamer), see TPR
TPF	Triphenyl phosphate [DIN, ISO], see TPP
TPP	Triphenyl phosphate [IUPAC], see TPF

(Table A-1, *cont.*)

TPR	1) 1,5-*trans*-poly(pentenamer), see TPA
	2) Thermoplastic elastomer, see TR
TR	Thermoplastic elastomer
TTP	Tricresyl phosphate, see TCF, TCP, TKP
UE	Polyurethane elastomer [ASTM]
UF	Urea-formaldehyde resin [ASTM, DIN, ISO]
UHMPE	Poly(ethylene) with ultrahigh molecular weight
UP	Unsaturated polyester [DIN]
UP-G-G	Prepreg of unsaturated polyesters and textile glass fabrics
UP-G-M	Prepreg of unsaturated polyesters and textile glass mats
UP-G-R	Prepreg of unsaturated polyesters and textile glass rovings
UR	Polyurethane elastomers [BS]
VA	Vinyl acetate
VAC	Vinyl acetate
VC	Vinyl chloride, see VCM
VC/E	Poly(ethylene-co-vinyl chloride)
VC/E/MA	Poly(ethylene-co-vinyl chloride-co-maleic anhydride)
VC/EV/AC	Poly(ethylene-co-vinyl chloride-co-vinyl acetate)
VCM	Vinyl chloride, see VC
VC/MA	Poly(vinyl chloride-co-maleic anhydride)
VC/OA	Poly(vinyl chloride-co-octyl acrylate)
VC/VDC	Poly(vinyl chloride-co-vinylidene chloride)
VF	Vulcan fiber
VPE	Cross-linked poly(ethylene)
VSI	Poly(dimethylsiloxane) with vinyl groups [ASTM]
WM	Plasticizer

TABLE A-2

Terms for general polymer properties and test methods used in publications quoted in this book.

Term	Other terms used	Test methods ASTM-D	DIN
Density	specific gravity	792, 941, 1550	51757, 53420, 53479
Color		1209	
Haze		1003	
Gloss		2457	
Transparency		1746	
Viscosity		1824	51550, 51562, 53015
Melt index	flow index, melt flow number	1238, 2116	53735
Surface tension		1331	
Water absorption		570	53471; 53473
Moisture regain			53728/1
Water vapor permeability	water vapor transmission rate	E96	53122; 53429
Gas permeability	gas permeation rate; gas permeability constant	1434	53380
Flammability		635	
Limiting oxygen index	LOI	2863	

TABLE A-3

Terms for mechanical properties.

Term	Other terms	Test methods ASTM-D	DIN
Tensile yield strength	yield strength; tensile strength at yield	638; 882; 1708; 412	53455
Yield	tensile yield; percentage elongation at yield	638; 882; 1708	53455
Modulus	tensile modulus; elastic modulus; modulus of elasticity in tension; initial modulus; Young's modulus	638; 882; 412	53455
Tensile strength	tensile; ultimate tensile strength; ultimate strength, burst strength, tensile strength at failure; tensile strength at fracture; tensile strength at break	638; 882; 1708	53455
Tensile strength at maximum load	tensile strength; tensile strength at maximum load; maximum strength		53455
Ultimate elongation	elongation; elongation at failure; percentage elongation; elongation at fracture; extensibility; breaking extension	1708; 638; 882; 412	53455

(Table A-3, *cont.*)

Term	Other terms	Test methods	
		ASTM-D	DIN
Tenacity			
Flexural strength	bending strength; flexural stress; flexural stiffness	790	53452 53423
Flexural modulus	flex modulus; modulus of flexional elasticity; bending modulus	790	53457
Compressive strength		695	53454 53421
Compression yield strength			53454
Compressive deformation	compressive strain		53454 53421
Compressive strain at compressive yield stress			53454
Compressive modulus		695	53454 53421
Shear strength		732	53422
Apparent modulus of rigidity	stiffness in torsion	1043	53447
Elastic shear modulus		2236	53445
Impact strength			53453
Impact strength with notch	Izod impact strength; notch Izod impact	256	53453
Impact tensile strength with notch			53448
Tear	tear resistivity; tear resistance; tear strength	470; 624	
Split tear		1938	53363
Creep modul	apparent modulus		53440
Folding endurance flexibility			
Flex life	flexing life		50100; 53574
Flex fatigue limit	flexing fatigue life; fatigue resistance	761	
Compression set		395	53572
Deformation under load		621	57275
Set at break			
Tensile elastic recovery			
Resilience	falling ball rebound	1054	53512
Loop strength	loop tenacity		
Knot strength			
Shore hardness	durometer hardness	2240 1706	53505
Rockwell hardness		785	53456
Abrasion	abrasion resistance	1044	

TABLE A-4
Terms for thermal properties.

Term	Other terms	Test methods	
		ASTM-D	DIN
Coefficient of linear thermal expansion		696	
Specific heat capacity	specific heat		
Thermal conductivity		C177	52612
Melt temperature	melting point	2117	
Glass transition temperature	glass temperature; second order transition temperature	846	
Brittleness temperature	brittleness temperature under impact; low temperature embrittlement	746	53372
Continuous service temperature			
Heat distortion temperature	heat distortion; heat resistance; heat deflection temperature; deflection temperature	648	53424 53461
Vicat softening temperature	Vicat softening point	1525	53460

TABLE A-5
Terms for electrical properties.

Term	Other terms	Test methods	
		ASTM	DIN
Dielectric constant	relative permittivity	150	57303
Dielectric strength	electric strength; breakdown strength	149	57303 53481
Dissipation factor	loss factor; dielectric loss factor	150	57303 53483
Dielectric loss index		150	53483
Volume resistivity	specific insulation resistivity; specific resistivity	257	57303 53482
Surface resistivity		257	57303 53482
Arc resistance		495	53484
Power factor	loss tangent	150	
Tracking resistance		2132	53480

TABLE A-6
SI units.

Physical quantity		Physical units	
Symbol	Name	Name	Symbol
Basic-physical quantities and units			
l	Length	meter	m
m	Mass	kilogram	kg
t	Time	second	s
I	Electric current	ampere	A
T	Thermodynamic temperature	kelvin	K
I_v	Luminous intensity	candela	cd
n	Amount of substance	mole	mol
Supplementary quantities and units			
$\alpha, \beta, \gamma \cdots$	Plane angle	radian	rad
ω, Ω	Solid angle	steradian	sr
Derived quantities and units			
F	Force	newton	$N = J\ m^{-1} = kg\ m\ s^{-2}$
E	Energy	joule	$J = N\ m = kg\ m^2\ s^{-2}$
P	Power	watt	$W = J\ s^{-1} = V\ A = kg\ m^2\ s^{-3}$
P	Pressure	pascal	$Pa = N\ m^{-2} = J\ m^{-3} = kg\ m^{-1}\ s^{-2}$
ν	Frequency	hertz	$Hz = s^{-1}$
Q	Electric charge	coulomb	$C = A\ s$
U	Electric potential difference	volt	$V = JC^{-1} = W\ A^{-1} = kg\ m^2\ s^{-3}\ A^{-1}$
R	Electric resistance	ohm	$\Omega = V\ A^{-1} = kg\ m^2\ s^{-3}\ A^{-2}$
G	Electric conductance	siemens	$S = A\ V^{-1} = s^3\ A^2\ kg^{-1}\ m^{-2}$
C	Electric capacitance	farad	$F = C\ V^{-1} = s^4\ A^2\ kg^{-1}\ m^{-2}$
ϕ	Magnetic flux	weber	$Wb = V\ s = kg\ m^2\ s^{-2}\ A^{-1}$
L	Inductance	henry	$H = V\ s\ A^{-1} = kg\ m^2\ s^{-2}\ A^{-2}$
B	Magnetic flux density	tesla	$T = V\ s\ m^{-2} = kg\ s^{-2}\ A^{-1}$
ϕ_v	Luminous flux	lumen	$lm = cd\ sr$
E_v	Illumination	lux	$lx = lm\ m^{-2} = cd\ sr\ m^{-2}$
ε	Relative permittivity	—	1

TABLE A-7
Prefixes for SI units.

Factor	Prefix	Symbol
10^{12}	tera	T
10^{9}	giga	G
10^{6}	mega	M
10^{3}	kilo	k
10^{2}	hecto	h
10^{1}	deca	da
10^{-1}	deci	d
10^{-2}	centi	c
10^{-3}	milli	m
10^{-6}	micro	μ
10^{-9}	nano	n
10^{-12}	pico	p
10^{-15}	femto	f
10^{-18}	atto	a

TABLE A-8
Fundamental constants.

Quantity	Symbol, value and unit(s)
Speed of light in vacuum	$c=(2.997\ 925\pm 0.000\ 003)\times 10^{8}$ m s^{-1}
Charge of proton	$e=(1.602\ 10\pm 0.000\ 07)\times 10^{-19}$ C
Faraday constant	$F=(9{,}648\ 70\pm 0.000\ 16)\times 10^{4}$ C mol^{-1}
Planck constant	$h=(6.625\ 6\pm 0.000\ 5)\times 10^{-34}$ J s
Boltzmann constant	$k=(1.380\ 54\pm 0.000\ 09)\times 10^{-23}$ J K^{-1}
Avogadro constant	$N_{A}=(6.022\ 52\pm 0.000\ 28)\times 10^{23}$ mol^{-1}
Gas constant	$R=(83.143\ 3\pm 0.004\ 4)$ bar cm^{3} K^{-1} mol^{-1}
	$=(8.314\ 33\pm 0.000\ 44)$ J K^{-1} mol^{-1}
Permeability of vacuum	$\mu_0=4\ \pi\times 10^{-7}$ J s^{2} C^{-2} m^{-1}
Permittivity of vacuum	$\varepsilon_0=\mu_0^{-1}\ c^{-2}=(8.854\ 185\pm 0.000\ 18)\times 10^{-12}$ J^{-1} C^{2} m^{-1}

TABLE A-9
Conversions of old units into SI units.

Old unit	= New unit
Lengths	
1 ft	= 30.48 cm = 0.3048 m
1 in.	= 2.54 cm = 0.0254 m
1 mil	= 0.002 54 cm = ~~0.002 54 m~~ 0.000 025 4 m
1 μ	= 1 μm = 10^{-6} m
1 mμ	= 1 nm = 10^{-9} m
1 Å	= 0.1 nm = 10^{-10} m

(Table A-9, *cont.*)

Old unit	= New unit
Areas	
1 sq. ft.	$=929.0304\ cm^2 = 9.290\ 304 \times 10^{-2}\ m^2$
1 sq. in.	$=6.4516\ cm^2 = 6.4516 \times 10^{-4}\ m^2$
Volumes	
1 cu. ft. (= 1. CF)	$=2.8317 \times 10^{-2}\ m^3$
1 gal (Brit)	$=4.5459\ dm^3 = 4.5459 \times 10^{-3}\ m^3$
1 gal (USA)	$=3.785\ dm^3 = 3.785 \times 10^{-3}\ m^3$
1 liter (old)	$=1.000\ 028\ dm^3 = 1.000\ 028 \times 10^{-3}\ m^3$
1 liter (new)	$\equiv 1\ dm^3 = 1 \times 10^{-3}\ m^3$
Masses	
1 lb	$=453.592\ 37\ g = 0.453\ 592\ 37\ kg$
1 short ton	$=907.184\ 74\ kg$
1 ton	$=1000\ kg$
1 long ton	$=1016.046\ 909\ kg$
Densities	
1 lb/cu. ft.	$=0.016\ 018\ 377\ g\ cm^{-3} = 16.018\ 377\ kg\ m^{-3}$
1 lb/cu. in.	$=27.679\ 905\ g\ cm^{-3} = 27\ 679.905\ kg\ m^{-3}$
Forces	
1 dyn	$=10^{-5}\ N = 10^{-5}\ kg\ m\ s^{-2}$
1 pond	$=9.806\ 65 \times 10^{-3}\ N = 9.806\ 65 \times 10^{-3}\ kg\ m\ s^{-2}$
1 kilogram-force	$=9.806\ 65\ N = 9.806\ 65\ kg\ m\ s^{-2}$
Pressures ($1\ N\ mm^{-2} = 1\ MN\ m^{-2} = 1\ MPa$)	
1 Phys. Atm.	$=101\ 325\ N\ m^{-2} = 101\ 325\ Pa = 101\ 325\ kg\ m^{-1}\ s^{-2}$
1 Techn. Atm.	$=98\ 065\ N\ m^{-2} = 98\ 065\ Pa = 98\ 065\ kg\ m^{-1}\ s^{-2}$
1 bar	$=10^5\ N\ m^{-2} = 10^5\ Pa = 10^5\ kg\ m^{-1}\ s^{-2}$
1 torr	$=(101\ 325/760)\ N\ m^{-2} = 133.3224\ Pa$
1 mmHg	$=133.322\ 387\ 4\ Pa$
$1\ kp/cm^2 = 1\ kgf/cm^2$	$=0.098\ 065\ MPa = 98\ 065\ kg\ m^{-1}\ s^{-2}$
1 lb/sq. in. = 1 psi	$=6.894\ 757 \times 10^{-3}\ N\ mm^{-2}$
$1\ dyn/cm^2$	$=10^{-7}\ N\ mm^{-2}$
Energies ($1\ J = 1\ kg\ m^2 s^{-2}$)	
1 erg	$=10^{-7}\ J = 10^{-7}\ kg\ m^2\ s^{-2}$
$1\ cal_{IT}$	$=4.1868\ J = 4.1868\ kg\ m^2\ s^{-2}$
$1\ cal_{th}$	$=4.184\ J$
1 eV	$=1.6021 \times 10^{-19}\ J$
1 kWh	$=3.6\ MJ$

Impact strengths

Impact strengths should be given as energy/area = force/length. Impact strengths with notch are also to be given in energy/area if related to the cross-section but carry the unit energy/length if they refer to the width of the notch.

1 kp/cm	$=9.806\ 65\ N\ cm^{-1} = 980.665\ kg\ s^{-2}$
$1\ kg\text{-}cm/cm^2$	$=9.806\ 65\ N\ cm^{-1} = 980.665\ kg\ s^{-2}$[a]
1 ft-lb/in. notch	$=53.378\ 659\ N = 53.378\ 659\ J/m$[a]

(Table A-9, *cont.*)

Old unit	= New unit
Tenacity	
Tenacities are really lengths:	
1 g/den	$= 9\ \mathrm{g\ tex^{-1}} = 0.9\ \mathrm{g\ dtex^{-1}} = 9\ \mathrm{km}$

Tenacities can be converted into true tensile strengths by multiplication with the densities:

$$\text{Tenacity} \times \text{density} = \text{tensile strength (in mass/area)}$$

The mass has, of course, to be converted into force.

Old unit	= New unit
Specific tenacity	
1 gf/den	$= 9\ \mathrm{cN/tex}$
Specific heat capacity	
1 BTU/lb. ft.	$= 4.187\ \mathrm{J\ g^{-1}\ K^{-1}}$
Heat conductivity	
1 BTU/ft h F	$= 6.234 \times 10^{3}\ \mathrm{J\ m^{-1}\ h^{-1}\ K^{-1}}$
Viscosities	
1 Poise	$= 0.1\ \mathrm{Pa\ s} = 0.1\ \mathrm{kg\ m^{-1}\ s^{-1}}$
1 Stokes	$= 10^{-4}\ \mathrm{m^{2}\ s^{-1}}$
Electric field strength	
1 V/mil	$= 0.3937\ \mathrm{kV\ cm^{-1}} = 39\ 370\ \mathrm{V\ m^{-1}}$
Permeabilities of gases and liquids	
1 g/sq. in./24 h/1 mil	$= 4.56 \times 10^{-11}\ \mathrm{g\ cm^{-1}\ s^{-1}}$
1 barrer	$= 7.50 \times 10^{-15}\ \mathrm{cm^{3}\ s\ g^{-1}}$
$\dfrac{1\ \mathrm{cm^{3}(STP)\ cm}}{\mathrm{cm^{2}\ s\ cmHg}}$	$= 7.50 \times 10^{-5}\ \mathrm{cm^{3}\ s\ g^{-1}}$
$\dfrac{1\ \mathrm{cm^{3}\ mil}}{100\ \mathrm{in.^{2}\ atm\ 24\ h}}$	$= 4.51 \times 10^{-17}\ \mathrm{cm^{3}\ s\ g^{-1}}$
$\dfrac{1\ \mathrm{ml}}{\mathrm{m^{2}\ h\ atm}}$	$= 2.75 \times 10^{-8}\ \mathrm{cm^{2}\ s\ g^{-1}}$
$\dfrac{1\ \mathrm{g}}{100\ \mathrm{in^{2}\ day}}$	$= 1.79 \times 10^{-8}\ \mathrm{g\ cm^{-2}\ s^{-1}}$
$\dfrac{1\ \mathrm{g\ mil}}{100\ \mathrm{sq.\ in.\ atm\ 24\ h}}$	$= 4.51 \times 10^{-17}\ \mathrm{s}$

The correct physical units are, if the permeating species is given in volume units,

$$\frac{\text{volume} \times \text{length}}{\text{area} \times \text{time} \times \text{pressure}} = \text{volume} \times \text{time/mass}$$

and, if the permeating species is given in mass units,

$$\frac{\text{mass} \times \text{length}}{\text{area} \times \text{time} \times \text{pressure}} = \text{time}$$

---

[a]Under substitution of masses of old units by forces of new units.

TABLE A-10
Addresses of corporations mentioned in this book.

Akzo NV
Velperweg 76
Arnhem, Netherlands

Allied Chemical Corporation
Specialty Chemicals Division
Columbia Road and Park Avenue
Morristown, NJ 07960, USA

American Cyanamid Company
859 Berdan Avenue
Wayne, NJ 07470, USA

American Enka Corporation
Enka, NC 28728, USA

American Kynol Incorporated
Carborundum Center
Niagara Falls, NY 14302, USA

American Smelting and Refining Co.
120 Broadway
New York, NY 10005, USA

Amoco Chemicals Corporation
Marketing Division
200 East Randolph Drive
P.O. Box 8640 A
Chicago, IL 60601, USA

Atlantic Research Corporation
Shorley Hwy. at Edsall Road
Alexandria, VA 22314, USA

BASF AG
D-67 Ludwigshafen am Rhein 1
Federal Republic of Germany

Bayer AG
Bayerwerk
D-509 Leverkusen
Federal Republic of Germany

Borg-Warner Corporation
200 S. Michigan Avenue
Chicago, IL 60604, USA

Carborundum Company
Engineering Plastics Group
P.O. Box 337
Bldg. 1-2
Niagara Falls, NY 14302, USA

Celanese Corporation
522 Fifth Avenue
New York, NY 10036, USA

Chemiefaser Lenzing AG
A-4860 Lenzing, Austria

Chemische Werke Hüls AG
D-437 Marl, Kreis Recklinghausen
Federal Republic of Germany

Ciba-Geigy AG
Division Kunststoffe und Additive
CH-4002 Basel
Switzerland

Dixon Corporation
386 Metacom Avenue
Bristol, RI 02809, USA

Dow Chemical Company
Midland, MI 48640, USA

DuPont Incorporated
1007 Market Street
Wilmington, DE 19898, USA

Dynamit-Nobel AG
Postfach 114-117
D-521 Troisdorf Bez. Köln
Federal Republic of Germany

Eastman Chemical Products
Plastics Division
P.O. Box 431
Kingsport, TN 37662, USA

(Table A-10, *cont.*)

Eastman Chemical Products International
Baarerstr. 8
CH-6301 Zug
Switzerland

Eastman Kodak Company
343 State Street
Rochester, NY 14650, USA

Esso Research and Engineering Co.
P.O. Box 111
Linden, NJ 07036, USA

Ferro Corporation
One Erieview
Cleveland, OH 44114, USA

Firestone Tire and Rubber Co.
1200 Firestone Parkway
Akron, OH 44317, USA

General Electric Company
570 Lexington Avenue
New York, NY 10022, USA

General Tire and Rubber Company
Chemical/Plastics Division
P.O. Box 951
Akron, OH 44329, USA

The P.D. George Company
5200 N. Second Street
St. Louis, MO 63147, USA

The B F Goodrich Company
500 S. Main Street
Akron, OH 44318, USA

The Goodyear Tire and Rubber Co.
1144 E. Market Street
Akron, OH 44316, USA

W. R. Grace and Company
Photopolymer Systems
Washington Research Center
7379 Route 32
Columbia, MD 21044, USA

W. R. Grace and Company
7 Hanover Square
New York, NY 10008, USA

Gulf Oil Chemicals Company
New Business Development Division
P.O. Box 2100
Houston, TX 77001, USA

Hercules Incorporated
910 Market Street
Wilmington, DE 19899, USA

Hexcel Corporation
11711 Dublin Blvd.
Dublin, CA 94566, USA

Hoechst AG
D-6230 Frankfurt (Main) 80
Federal Republic of Germany

Hughes Aircraft Company
Centinela Ave. and Teale Street
Culver City, CA 90230, USA

Imperial Chemical Industries Inc.
Wilmington, DE 19899, USA

Imperial Chemical Industries Ltd.
Plastics Division
P.O. Box 6
Bessmer Road
Welwyn Garden City, Hertford, England

Japan Synthetic Rubber Company
No. 1, 1-chome, Kyobashi
Chuo-ku, Tokyo, Japan

Japan Synthetic Rubber Company
Grove House
551 London Road
Isleworth, Middlesex, England

(Table A-10, *cont.*)

Japan Synthetic Rubber Company
Empire State Building
Suite 8001
350 Fifth Avenue
New York, NY 10001, USA

Kelco Co. subsidiary of Merck and Co., Inc.
126 E. Lincoln Avenue
Rahway, NJ 07065, USA

Kohjiin International
441 Lexington Avenue
New York, NY 10017, USA

Lithium Corporation
2 Pennsylvania Plaza
New York, NY 10001, USA

LONZA-Werke GmbH
D-7858 Weil am Rhein
Federal Republic of Germany

Marine Commodities International
Shrimp Harbor
Star Route Box 140
Brownsville, TX 78520, USA

Minnesota Mining and Mfg. Co.
3M Center
St. Paul, MN 55101, USA

Mobay Chemical Company
Penn Lincoln Parkway, W.
Pittsburgh, PA 15205, USA

Monsanto Company
800 N. Lindbergh Blvd.
St. Louis, MO 63166, USA

Montedison
Foro Buonaparte 31
Milano, Italy

Olin Corporation
120 Long Ridge Road
Stamford, CT 06904, USA

PCR Incorporated
Box 1466
Gainesville, FL 32601, USA

Phillips Petroleum Company
Chemical Department
Plastics Division
Bartlesville, OK 74004, USA

Polymer Corporation
2120 Fairmont Avenue
Reading, PA 19603, USA

Radiation Research Corporation
1150 Shames Drive
Westbury, NY 11590, USA

Raychem Corporation
300 Constitution Drive
Menlo Park, CA 94025, USA

Rhône-Poulenc SA
21 rue Jean-Goujon
Paris-8^e, France

Richardson Company
2700 W. Lake Street
Melrose Park, IL 60160, USA

Röhm GmbH
Postfach 4166
Kirschenallee
D-6100 Darmstadt
Federal Republic of Germany

Royal Dutch/Shell
Carel van Bylandtlaan 30
The Hague, Netherlands

Schenectady Chemicals, Inc.
P.O. Box 1046
Schenectady, NY 12301, USA

(Table A-10, *cont.*)

Scott Paper Company
Scott Plaza
Philadelphia, PA 19113

Shell Chemical Company
50 W. 50th Street
New York, NY 10020, USA

Solvay et Compagnie
33 rue Prince Albert
Brussels, Belgium

Standard Oil Company
Midland Bldg.
Cleveland, OH 44115, USA

Stauffer Chemical Company
299 Park Avenue
New York, NY 10017, USA

Stein, Hall and Co., Inc.
605 Third Avenue
New York, NY 10016, USA

Thiokol Chemical Corporation
Newportville Road
P.O. Box 27
Bristol, PA 19007, USA

Union Carbide Corporation
270 Park Avenue
New York, NY 10017, USA

Unitika Ltd., Plastics Division
4-68, Kitakyutaro-Machi, Higashi-ku
Osaka, Japan

Upjohn Polymer Chemicals
Box 685
La Porte, TX 77571, USA

Wacker-Chemie GmbH
Prinzregentenstr. 22
D-8000 München 22
Federal Republic of Germany

Subject Index